CONFERENCE
PROCEEDINGS

RAND Forum on Hydrogen Technology and Policy

A Conference Report

Mark A. Bernstein

INFRASTRUCTURE, SAFETY, AND ENVIRONMENT

The research described in this report was funded by a consortium of companies and institutions interested in hydrogen technology and uses.

0-8330-3817-6

Published 2005 by the RAND Corporation
1776 Main Street, P.O. Box 2138, Santa Monica, CA 90407-2138
1200 South Hayes Street, Arlington, VA 22202-5050
201 North Craig Street, Suite 202, Pittsburgh, PA 15213-1516
RAND URL: http://www.rand.org/
To order RAND documents or to obtain additional information, contact
Distribution Services: Telephone: (310) 451-7002;
Fax: (310) 451-6915; Email: order@rand.org

Preface

In recent years, hydrogen has drawn much attention due to its potential large-scale use in producing electrical energy through stationary fuel-cell technologies and its potential for replacing gasoline for use in transportation. Among the advantages of hydrogen are its abundance and its ability to produce electricity in some applications with virtually no harmful emissions. Among its disadvantages are that it cannot be used without being transformed through a series of processes that require significant energy input.

On December 9, 2004, the RAND Corporation hosted a forum on hydrogen energy that drew 40 experts in various fields from the United States, Canada, and Norway. The goal of the forum was to facilitate an open discussion on the analyses and actions that are needed to inform decisionmakers in the public and private sectors on the opportunities, benefits, and costs of various hydrogen-related programs and policies.

The forum participants represented a number of public and private organizations. They had varied interests in as well as varied perspectives on the future of hydrogen as an alternative energy carrier. The participants included energy consultants and members of California and federal government agencies, private-sector companies, universities, and RAND. While not every participant expressed optimism about the use of hydrogen in the near term, almost all are invested in hydrogen technology in some way and most have the belief that, at some time in the future, hydrogen can be used as an energy carrier.

This report summarizes the forum proceedings. The forum was conducted on a not-for-attribution basis to encourage candor from participants. The views expressed in this document are those of the participants, as interpreted by the RAND Corporation, and do not represent RAND analysis. This report should be of interest to individuals in the policy, business, and research communities who are involved in hydrogen production, distribution, and applications and those who are interested in energy issues in general.

This research was conducted within RAND Infrastructure, Safety, and Environment (ISE), a unit of the RAND Corporation. The mission of ISE is to improve the development, operation, use, and protection of society's essential built and natural assets, and to enhance the related social assets of safety and security of individuals in transit and in their workplaces and communities. The ISE research portfolio encompasses research and analysis on a broad range of policy areas including homeland security, criminal justice, public safety, occupational safety, the environment, energy, natural resources, climate, agriculture, economic development, transportation, information and telecommunications technologies, space exploration, and other aspects of science and technology policy.

Inquiries regarding RAND Infrastructure, Safety, and Environment may be directed to:

Debra Knopman, Director
1200 S. Hayes Street
Arlington, VA 22202-5050
Tel: 703.413.1100, extension 5667
Email: ise@rand.org
http://www.rand.org/ise

Contents

Summary

In recent years, hydrogen has drawn much attention due to its potential large-scale use in producing electrical energy through stationary fuel-cell technologies and in replacing gasoline for use in transportation. Among the advantages of hydrogen are its abundance and its ability to produce electricity in some applications with virtually no harmful emissions. Among its disadvantages are that it cannot be used without being transformed through a series of processes that require a significant energy input.

Decisionmakers in the public and private sectors do not have all the information they need for determining whether to invest in hydrogen research or to make investments in the infrastructure that would be needed to use hydrogen as a source of energy. Decisionmakers also lack information to help them decide whether to formulate policies that will hasten the development of hydrogen as a viable energy source.

This report provides an overview of the discussions that took place during a daylong forum on December 9, 2004, that was hosted and organized by the RAND Corporation. The forum was intended to facilitate open discussion of issues related to making hydrogen a viable alternative energy source and to describe a set of analyses and actions that are needed in the public and private sectors to improve decisionmaking on investments in hydrogen. The forum was in the format of a facilitated discussion. Each session of the forum started with a stated goal for the session or a question or anecdote to prompt discussion, and the floor was then opened for dialogue.

Potential Benefits of Hydrogen for Further Evaluation

A major conclusion drawn by forum participants was that while studies have been done on hydrogen technology, and policy papers have discussed numerous possible benefits that might accrue from the introduction of hydrogen as an energy carrier, some benefits of hydrogen have not been adequately addressed either in quantitative analyses or in policy discussions. (Hydrogen is referred to as an *energy carrier* because, like electricity, it needs to be made from a primary energy source, such as natural gas.)

The forum discussion was framed in the context of whether private-sector companies or the government should make investments in hydrogen research, development, and deployment. While forum participants did not address the costs of hydrogen, they identified the following potential benefits of hydrogen, which warrant further examination and assessment:

- Introducing hydrogen as an alternative energy source could add diversity to the supply of transportation fuels, thereby making the United States less dependent on petroleum and making fuel costs more stable and predictable.
- If hydrogen-based fuel cells were put to use generating electricity on a small scale close to areas where electricity is needed, the burden on the current electric grid—the system that generates and distributes electricity—could be eased.
- If renewable energy is used to make hydrogen, fuel cells could provide a means of storing renewable electricity—something that cannot be done today.
- If communities and companies had the ability to generate their own electricity via small fuel cells using renewable energy to make hydrogen, they could fulfill their energy needs locally and would not have to depend as much on imported energy.
- Private companies that develop innovative technologies for using hydrogen as an alternative energy source have the potential to become highly profitable, world-class technology leaders.
- Developing nations that put hydrogen to work right away could leapfrog over the environmentally destructive practices that have occurred in other countries.
- Reducing the use of petroleum could also reduce the environmental impacts of exploring for, producing, transporting, and refining petroleum, including the potential contamination of groundwater and surface water.

Risks of Inaction Perceived as Being Substantial

In addition to the benefits that might accrue from making investments in hydrogen, the participants concluded that there are significant risks in *not* making investments in hydrogen. While the participants pointed out that there are risks in making too large an investment too quickly, they believed that the risks from no action are greater than those from some action for various scenarios of the future. The group cited risks to the environment (both locally, in terms of pollution, and globally, in terms of climate change) as the most significant risks, followed by economic risks, of not taking actions to invest in hydrogen. These risks derive from the increasing costs associated with mitigating growing environmental problems, but also from the possibility that other countries will take the technological lead in hydrogen and renewable technologies, causing U.S. companies to lose economically. Additional risks include dependence on a single source of energy for transportation and risks from potentially reduced reliability of the electricity supply.

Hurdles to Implementing Hydrogen as an Energy Carrier

The discussions among forum participants frequently returned to the subject of the need to understand basic hydrogen infrastructure issues. That is, what will it take to make hydrogen work as an energy carrier or source of electricity? While the group acknowledged that there were technology hurdles to cross before hydrogen could be implemented as a transportation or electricity energy source, the general feeling among the group was that those hurdles could be overcome and that it would not take very long to do so. On the other hand, some other significant issues were identified that may not be so easily addressed:

- The question of who is going to pay for the hydrogen development activity that needs to occur between the research phase (which might be funded primarily by the government) and commercial deployment (which would consist of investments by the private sector)
- Lack of a coherent energy policy, which will hinder investments in hydrogen
- Regulatory roadblocks to introducing hydrogen
- Perception problems with hydrogen—primarily regarding the safety of hydrogen (on the part of the public) and regarding market opportunities (on the part of the private sector)
- Lack of a consistent set of economic metrics to value hydrogen that are needed to produce robust cost-benefit estimates.

Going Forward

When decisions concerning major technological transitions are on the horizon, they can often be informed by lessons learned during similar transitions in the past. Participants cited lessons to be learned from past efforts to ramp up biomass fuel programs (the use of organic matter to produce heat energy) and natural gas fuel programs, but also noted that the transition to hydrogen may substantially differ from those earlier experiences. Participants discussed the possibility that lessons may be learned from technological transitions in other markets—e.g., computers, compact disks, and MP3 players. Technology-diffusion paradigms may be shifting, participants observed, and technical specialists and decisionmakers need to incorporate these new paradigms in their assessments of how a transition to hydrogen might occur.

A consistent message from forum participants expressing a public-policy point of view was that hydrogen as an energy source could provide substantial benefits for California and for the United States as a whole. Participants said that more information is needed to help policymakers determine what role the government should, or should not, play in furthering the development of hydrogen. The U.S. Department of Energy's Hydrogen Posture Plan and the California Hydrogen Highway Blueprint Plan are both good jumping-off points for the development of hydrogen, but participants pointed out that the transition to hydrogen will not happen unless more robust, more objective, and more transparent information is made available to public- and private-sector decisionmakers. There is clearly a role that the public sector can play in assisting in the development of this information.

The private sector needs to better understand the prospects for hydrogen energy and the value of investments in hydrogen, and its investment decisions need to reflect an understanding of the risks associated with current patterns of energy use. Participants said that it is critically important for companies that are already engaged in the development of hydrogen-use technologies to demonstrate that the technologies are reliable and that they have the ability to warranty their "product," thereby reassuring the financial community of the viability of hydrogen.

There seemed to be general agreement that sooner is better than later for the public and private sectors to invest in hydrogen as an energy carrier. While there were differing opinions on how large the hydrogen energy market would be today, the general opinion was that sufficient technological improvements have been made in the past few years to make the hydrogen energy marketplace viable for commercial development. However, the development of hydrogen energy needs a boost from government, and policymakers still need convincing to move aggressively forward on hydrogen policy, participants observed. Policymakers need more information on the unique potential benefits of hydrogen, the new opportunities for investments and jobs, and how a portfolio of policies and investment options can meet short-term and long-term goals for policy actions. While hydrogen as an energy carrier is not the only new technological and market opportunity available to investors, participants said that hydrogen, nevertheless, should be a significant part of the U.S. public and private investment portfolio.

Acknowledgments

RAND would like to thank the following organizations for their generous support of the forum and of this document: Toyota; South Coast Air Quality Management District; California Air Resources Board; Air Products and Chemicals, Inc.; Bechtel; Chevron Corp.; Parsons Corporation; Ballard Power Systems, Inc.; Gas Technology Institute; Stuart Energy; Arizona Public Service; Applied Research and Engineering Services; IF, LLC; and Nuvera Fuel Cells.

Special thanks also go to David Haberman from IF, LLC, for encouraging participation in the forum. I also want to express appreciation to a number of individuals at RAND: Sergej Mahnovski, who helped coordinate notes for the forum and contributed to the background sections of this report; Aaron Kofner and Jay Griffin, who took notes during the forum; Lloyd Dixon, Rob Lempert, and D. J. Peterson for their comments on the drafts of this report; Shelley Wiseman for helping to shape the report; and Nancy DelFavero for a fantastic editing job.

CHAPTER ONE

Introduction

In recent years, hydrogen as an energy carrier[1] has generated much enthusiasm and discussion among policymakers and industry over its potential large-scale use in stationary fuel-cell technologies to produce electrical energy and in fuel-cell powered cars. Hydrogen is the world's most abundant chemical element and is already used in various industrial applications. Among the commonly cited advantages of hydrogen as an energy carrier are its abundance and its ability to produce electricity in some applications with virtually no harmful emissions. Among the oft-cited disadvantages are that it is not a primary energy source, and it cannot be used without being transformed or "produced" by a series of processes that require a significant input of energy. Despite active research programs, fuel cell and hydrogen conversion and storage technologies still have not been perfected; therefore, hydrogen energy remains more expensive than energy produced with conventional fuel sources such as oil, coal, and hydroelectric power and alternative energy sources such as wind and solar power.

The public and private sectors are actively exploring hydrogen's potential as an energy carrier. However, it is also understood among those who are have an interest in hydrogen-energy issues that the analyses that have been conducted to date of the benefits, barriers, risks, and costs related to the development of hydrogen as an energy source are not necessarily conclusive; rather, they provide a basis upon which new tools can be developed for conducting robust analyses to guide decisionmaking regarding investments in hydrogen technology. In many ways, the uncertainty surrounding the future of hydrogen is representative of the challenges and pitfalls of long-term technology and energy forecasting and analysis in general (see the related discussion under "Forecasting the Future Is Not Simple: A Cautionary Tale").

RAND Forum Goals and Forum Participants

On December 9, 2004, the RAND Corporation hosted a forum on issues related to the development of hydrogen as an energy source. The goals of the RAND forum were to facilitate an open discussion of the opportunities and challenges associated with promoting hydrogen as an energy source and to describe a set of analyses and actions that are needed in the public and private sectors to improve decisionmaking about investments in hydrogen. The discussions took place at a time when the State of California was preparing a blueprint for its

[1] The term *energy carrier* refers to hydrogen's having to be produced (e.g., electricity is an energy carrier) rather than being an energy *source* (e.g., oil, which is found in nature, is a primary source of fuel).

Hydrogen Highways program and the U.S. government was completing its Hydrogen Posture Plan.

The 40 forum participants brought to the table their varied experience and perspectives on the future of hydrogen. They represented companies involved in the research and development of applications of hydrogen and production of hydrogen, universities conducting analyses of hydrogen, organizations responsible for implementing policy that could impact the use of hydrogen, and researchers from the RAND Corporation. (See Appendix D for a list of forum participants and their affiliations.)

It should be noted here that most of the forum participants are invested in hydrogen in that their organizations are making significant financial investments in hydrogen research or are creating products for a future in which hydrogen is a significant energy source, or they are involved in developing policy issues in which hydrogen may play a significant role. So while the participants were cautious about the future of hydrogen, and there were disagreements among them about the extent to which hydrogen will be used and how quickly it will become part of the energy portfolio, most of those in the group foresaw a significant future for hydrogen as an energy carrier.

Forecasting the Future Is Not Simple: A Cautionary Tale

In 1963, Resources for the Future, a nonpartisan organization that conducts research on environmental and natural resource issues, published the first real forecast of resource use for the United States (Landsberg, Fischman, and Fisher, 1963). It was groundbreaking work that shaped the way energy analysis has been done for more than 40 years. Twenty-two years after the publication of the report, one of the report's authors, Hans Landsberg, looked back at the work and compared what the analysis had forecasted for 1985 with what actually happened. (He presented his findings in a number of lectures and in Landsberg [1985)]). Much happened between 1963 and 1985 that was clearly not anticipated (for example, the enactment of environmental policies such as the Clean Air Act and the oil embargoes of the late 1970s that caused rising oil prices that led to improvements in energy efficiency). These events clearly had an impact on energy use. Even so, the original forecast for total energy use in the United States for 1985 was remarkably close to the actual energy that was used. However, almost all of the underlying assumptions were not very accurate. So while the sum of the pieces was prescient, the pieces themselves were not. The caution from this exercise is that even if we can outline the critical hydrogen technology issues for analysis, we need to acknowledge that our ability to forecast the future is limited, and uncertainty will continue to exist even if we believe that we have done the best analysis possible.

References: Landsberg, Hans H., "Energy in Transition: A View from 1960," *The Energy Journal,* Vol. 6, No. 2, April 1–18, 1985; Landsberg, Hans H., Leonard L. Fischman, and Joseph L. Fisher, *Resources in America's Future: Patterns of Requirements and Availabilities, 1960–2000,* Baltimore, Md.: Johns Hopkins University Press, 1963.

About This Report

The organization of this report roughly follows the order of the topics listed in the forum's agenda (see Appendix C.) Chapter Two summarizes the forum's opening discussion of benefits that may result from investments in hydrogen technology and a description of the anticipated timeframes over which these benefits can be achieved. Chapter Three reviews the group's discussion of barriers to the implementation of hydrogen, which included a lengthy discourse on the "valley of death" for technology innovation—i.e., the funding gap that lies between the research and development stage and commercial viability. Chapter Four addresses the risks of various policy approaches to promoting hydrogen technology. Chapter Five summarizes the additional information that participants said they would need if they had to make a case for or against investments in hydrogen.

Each chapter also includes supplementary sidebar information on issues that were covered in the course of the forum discussions. The sidebar material represents both anecdotal information used as lead-ins to forum discussions and synopses of related literature that was mentioned during the discussions or that was included in this report at the suggestion of participants.

Appendix A provides background information on hydrogen, including what it is, how it is produced and used, how it might be used in the future, and technological hurdles to achieving hydrogen-energy applications. Appendix B lists the potential benefits of hydrogen and the potential barriers to the development of hydrogen technology that were cited by forum participants during brainstorming sessions. Appendix C lists the forum agenda, and Appendix D lists the individuals and organizations represented at the forum. Finally, Appendix F presents matrices displaying the impact of three approaches to hydrogen policy: market-only, moderate, and aggressive. The matrices display the level of impact for various investment and policy goals given several future scenarios.

CHAPTER TWO

Public-Sector and Private-Sector Benefits of Investing in Hydrogen

The goal of the forum's first facilitated discussion was to elicit from participants a description of the benefits that could accrue to public- and private-sector investors if hydrogen were fully developed as an alternative energy source, assuming of course that certain technological hurdles are overcome. (For a discussion of those hurdles, see Appendix A.) This discussion preceded the discussion of barriers to developing hydrogen as an energy carrier (see Chapter Three) and was not intended to be encumbered by practical considerations; nor was the intention to have participants report on proven benefits that are supported by analysis. Rather, this portion of the forum was intended to be a wide-open brainstorming session about hydrogen's potential benefits and why participants believe that the government and the private sector should consider investing in hydrogen.

Participants did not address the costs associated with a transition to hydrogen because many of them felt that the cost side of such a transition was relatively well known and understood. As such, this chapter is limited to recounting some key benefits cited by participants, and especially benefits that participants felt are underrepresented in analyses. (For a complete list of the benefits cited by participants, see Appendix B.) Concluding this chapter is a summary of the group's input on the optimum timeframe for accruing benefits from hydrogen that would be necessary to make development of hydrogen technology viable.

Social Benefits from Government Investment in Hydrogen

Participants cited three general categories of potential benefits that may accrue to the public should governments choose to invest significant resources to promote hydrogen production and distribution and hydrogen's use as an energy source:

- Reduction in the consumption of oil in the transportation sector
- Improvement in the efficiency and reliability of the electric grid
- Reduction of other environmental problems that are not attributable directly to oil consumption.

Reduction in the Consumption of Oil

If hydrogen becomes a reliable source of energy for cars and other modes of transportation, the overall impact in the United States could be a reduction in the consumption of oil. Participants observed that a reduction in oil consumption could result in a number of benefits:

- Providing diversity in the mix of transport fuels, ensuring a steady supply of transport fuels, giving consumers more choices on fuels, and making transportation costs more predictable
- Reducing U.S. dependence on foreign oil
- Reducing the chances of financially and environmentally costly oil spills
- Improving air quality.

The reduction in oil use and the introduction of hydrogen as an energy carrier can provide diversity in transport fuels. Currently, U.S. transportation is about 95 percent dependent on oil, and there is little excess fuel capacity in the U.S. transport system, particularly in refineries in the United States. This can lead to uncertainty and volatility in fuel prices, and the only way that consumers can hedge against fluctuating prices is to use less fuel. There are not many options for oil suppliers to hedge prices either. However, a diverse set of fuels can provide ways to hedge transportation costs and make them more predictable.

Reductions in oil use can also have implications on U.S. foreign policy as it relates to oil-exporting nations, according to forum participants. The United States imports more than 50 percent of the oil it needs. If the reduction in oil consumption leads to a reduction in oil imports, some foreign policy actions, which are partially driven by concerns over oil supplies, might be ameliorated. Further, these reductions in imports can reduce the U.S. foreign trade deficit of which the share of oil is becoming an increasingly large portion. If hydrogen displaces oil, it will likely displace the most expensive oil first, which could be domestically produced oil.

It was noted during the forum that less oil use could reduce the chance of oil spills that can contaminate water sources, including surface and groundwater sources, as well as the oceans. Recent oil spills off the coasts of Europe and the United States and increasing evidence of oil-related products leaching into drinking water highlight the problems associated with oil use. There are technologies that can reduce the probability of oil contamination in the environment, but these technologies would not eliminate the possibility of contamination altogether and could lead to higher prices, which, in turn, could make alternatives like hydrogen more attractive.

Finally, depending on how and where hydrogen is produced, reductions in oil use for transportation can have a significant impact on urban air pollution and in particular on ground-level ozone and particulates, which continue to be significant problems for many regions of the United States (see the accompanying discussion under "Problems with Ground-Level Ozone").

Improving the Efficiency and Reliability of the Electric Grid

The second major category of benefits highlighted by forum participants is associated with electricity generation.

If hydrogen-powered fuel cells can be used for small-scale electricity generation, and if technologies to produce hydrogen improve such that hydrogen can be delivered efficiently and cheaply to those small-scale generators, there are possible benefits to the transmission and distribution system (these small-scale generators can use other fuels such as natural gas).

Problems with Ground-Level Ozone

Air pollution continues to be a problem in the United States despite the considerable progress that has been made over the past 30 years toward meeting clean air goals. With regard to automobile transportation, there are two key emissions of concern—nitrous oxides (NOx) and volatile organic compounds (VOCs). NOx and VOCs are key ingredients in the formation of ground-level ozone, which presents well-recognized health and environmental hazards. Many parts of the United States have experienced unhealthy air because of high concentrations of ozone, even though almost all geographic areas of the country have made progress in lowering their emissions of pollutants that are precursors to ozone. In 2002 in the United States, the annual number of days in which ozone levels were deemed to be unhealthy was nine higher (or more than 20 percent higher) than the average annual number of such days between 1998 and 2001. As of July 15, 2003, the number of unhealthy ozone-level days was already twice the number observed at that point in 2002 (Polakovic, 2003).

One-third of the U.S. population faces a risk of health effects related to ground-level ozone. Children, for example, are at greater risk of respiratory problems because they generally engage in more outdoor activities than adults and because their lungs are still developing. Individuals with existing respiratory problems are also at greater risk. A study of 271 asthmatic children in southern New England, reported in the *Journal of the American Medical Association* (JAMA), found that even ozone levels that fell within air quality standards set by the Environmental Protection Agency affected the severity of the children's asthma (Bell at al., 2004). These results are consistent with previous studies cited in the JAMA article that found that even with low levels of ambient ozone and controlling for the presence of fine particulate matter, children with severe asthma have a high risk of experiencing respiratory symptoms from ground-level ozone.

References: Bell, Michelle L., Aidan McDermott, Scott L. Zeger, Jonathan M. Samet, Francesca Dominici, et al., "Ozone and Short-Term Mortality in 95 U.S. Urban Communities, 1987–2000," *Journal of the American Medical Association,* Vol. 292, No. 19, November 17, 2004, pp. 2372–2378; Polakovic, Gary, "Smog Woes Back on Horizon," *Los Angeles Times,* July 15, 2003, p. A1.

Locating power sources closer to where electricity is used puts less strain on the electricity transmission and distribution lines. It is increasingly difficult and expensive to site and build new power lines, so if the old lines are nearing capacity, "load-centered generation" can postpone the need to build new lines and reduce the chance of power outages (see the discussion under "Benefits of Load-Centered Generation").

Participants pointed out that hydrogen-powered fuel cells might also complement renewable energy sources such as photovoltaics (PVs) (solar cells that absorb sunlight and convert it directly into electricity). The main problem with PVs is that they need sunlight and cannot generate power at night or on overcast days. Some PV installations have used batteries as supplementary power sources, but batteries are relatively inefficient and expensive. On the other hand, if some of the PV power is used as the needed power source to create hydrogen during the daytime, it may be possible that the fuel cell could be used at night when the PV is not producing electricity, thereby providing "storable" renewable energy (research in this area is ongoing at the National Renewable Energy Lab). Some technology improvements need to occur, participants observed, particularly in hydrogen storage efficiency, to make this "storable" renewable energy viable, but the opportunity to create storable energy can result in a key long-term benefit of using hydrogen. The complement of PV and hydrogen also provides a potential benefit for remote power applications. If the efficiency of electrolysis (the process by which water is separated into hydrogen and oxygen) improves, a hybrid system composed of PV and a hydrogen-powered fuel cell could be run nearly anywhere, assuming there is the necessary water for the electrolysis process, thus providing power in an isolated, remote setting.

Reducing Environmental Problems

The third general category of benefits mentioned by participants relates to the environment (beyond the environmental benefits specifically associated with reducing petroleum use).

Benefits of Load-Centered Generation

Load-centered generation refers to the practice of generating electricity as close as possible to areas where there is the most demand for it, thereby reducing the need to send the electricity long distances and reducing the reliance on the system of overhead and underground wires that make up the U.S. transmission grid. Much of California's grid of 26,000 miles of transmission lines is operating under great strain. It is part of the 115,000-mile western grid that stretches from British Columbia to northern Mexico, linking more than 700 power plants. Several major transmission corridors operate close to their capacity, including the widely publicized Path 15, which links Northern and Southern California.

In January 2001, Northern California, which was unable to secure its accustomed electricity supply from the drought-stricken Pacific Northwest hydroelectric plants, suffered rolling blackouts when excess capacity in Southern California could not be transmitted through Path 15.

An overstrained transmission grid is vulnerable to a loss of service at any location; for example, in early April 2001, a windstorm knocked out a transmission line between the Northwest and Southern California, depriving Los Angeles of 3,000 megawatts of transmission capacity for ten days and causing a Stage 2 emergency.

Load-centered generation relieves much of the strain on the transmission grid imposed by central-station generation and allows utility planners to defer transmission-line investments. Estimates of the savings from these deferred investments range from about one cent to seven cents per kilowatt hour.

Reference: Bernstein, Mark, Paul Dreyer, Mark Hanson, and Jonathan Kulick, *Load-Centered Power Generation in Burbank, Glendale, and Pasadena: Potential Benefits for the Cities and for California,* Santa Monica, Calif.: RAND Corporation, IP-214-BGP, 2001.

These benefits are primarily associated with the potential to reduce greenhouse gas emissions, and they critically depend on how hydrogen is produced. If hydrogen is produced through non–carbon-intensive sources, then there can be a net reduction in greenhouse gas emissions.

A forum participant who is a representative of the energy industry initiated the discussion, saying that, "Carbon sequestration is something that we're trying to accomplish. One of the big contributors is coal, an enormously abundant resource. The DOE [U.S. Department of Energy] spent a lot of money chasing synthetic methane. Can hydrogen play a role in creating synthetic methane, which would have an immediate impact on production of CO_2 on a global basis? Could methane then be used as a vehicle fuel? Why was the DOE's vision from a generation ago aborted? Why does hydrogen have such momentum today?"

On the other hand, some participants countered, if advances occur in the ability to sequester carbon (store it in a form that will not migrate to the atmosphere), it would still be possible to use carbon-rich energy sources such as coal to produce hydrogen and gain environmental benefits. Carbon dioxide is one of the potentially harmful byproducts that result from producing hydrogen when using energy sources such as coal. The assumption is that it will be easier and more cost effective to sequester carbon in large-scale facilities and less likely that carbon sequestration will be possible in smaller settings or "on the fly" in mobile applications such as cars. Hydrogen could be produced using coal at large, centrally located facilities that are equipped to sequester the carbon that results from the process. In this scenario, the hydrogen fuel would be produced in a way that minimizes emissions of greenhouse gases, and it could then be distributed or applied to mobile applications.

Other Public Benefits

One participant, a representative from the energy industry, noted that there is a "tremendous amount of worry and a sense of there being problems in the world related to oil in the Middle East and personal security. [The potential for hydrogen to help] reduce tensions and ameliorate foreign policy problems could benefit people's sense of well being."

Participants offered other examples of benefits: Hydrogen technologies could also provide opportunities for developing nations to take more control over their energy sources (relying more on their own sources rather than on international ones) and provide electrical services to rural areas where almost two billion people now have no access to electricity. Hydrogen technologies could allow these countries to provide more energy to their citizens with less impact on the environment than the impact that has occurred in industrialized nations. In one scenario posited by a participant, micro-grid applications in remote villages might allow local water supplies to be used with PV, wind, and/or biomass (organic matter) energy to accomplish two goals—make use of water supplies to convert the hydrogen for energy and at the same time clean the water for human consumption. As such, micro-grid applications can be an efficient and effective option for remote locations.

Finally, participants mentioned the potential for spin-off technologies and applications. For example, advances in membrane technologies for fuel cells may have medical applications. Other spin-offs could occur, and while it is not possible to quantify these benefits now, the potential opportunities from spin-offs could be great.

Private-Sector Benefits

Forum participants felt that it was important to discuss the benefits that can accrue from investments in hydrogen technology by private-sector companies and that, in general, those benefits are overlooked in cost-benefit analyses that tend to focus on social benefits. The discussion focused on why companies might choose to invest in the early stages of hydrogen development and deployment, as well as investing in the later stages when the technology is commercialized.

One industry analyst noted that in some areas hospitals are looking to use distributed generation for a "pure electrical supply, particularly in applications where reliability of energy supply is crucial."

Using hydrogen as an energy source could reduce a company's environmental liabilities in the future. In particular, if companies were to use some hydrogen today to replace oil as transportation fuel or to replace coal in coal-based electricity, and if they are able to reduce pollution, they will also reduce their potential future liabilities associated with that pollution. For example, it was noted that after the market for emissions credits related to greenhouse gases is established by the U.S. Environmental Protection Agency and other regulatory agencies, companies may find that their operations are running so cleanly that they have emissions credits they can sell.

For companies that require a lot of energy to operate, it was noted during the session, investments in hydrogen could give those companies more control over their energy sources, make their energy portfolio more diverse and, therefore, make their costs more predictable, or at least make it easier for them to hedge against rising prices. In the near term, if companies generate their own power (whether using fuels cells or other sources), they will reduce their demands on the larger electric grid during times of peak demand and have a significant impact on reducing their electricity costs, because peak-demand charges in some regions of the country are quite high.

Finally, there may well be profits for companies creating hydrogen-based technologies; U.S. companies may find themselves on the leading edge of a world-class industrial base.

Other Technologies That Can Provide Similar Benefits

An important counterpoint that was made during the session was that other technologies could provide public- and private-sector benefits similar to those attributed to hydrogen but perhaps with fewer dollars of investment than hydrogen requires. For example, hydrogen is not the only means for reducing oil consumption in the transportation sector. Other alternative fuels that could probably be produced less expensively include natural gas and biomass-based fuels (e.g., ethanol), although previous attempts to significantly increase the use of these alternatives have not been successful. More-efficient vehicles, including electric-hybrid vehicles, can also reduce oil consumption. Technologies are available now that can make new vehicles cleaner and more efficient, and increased public transportation and sustainable land-use planning can have a significant impact on future emissions.

There are other options for generating small amounts of electricity locally, including micro-turbines fueled by natural gas, diesel engines, and PVs. Micro-turbines can also help provide more reliable power sources for private companies that want to take more control of their energy needs. The additional benefit from hydrogen in this application is that it produces no pollution. In areas of the country that already do not meet air-quality goals, it may not be possible to introduce micro-turbines and generators, which produce some levels of pollution.

Participants pointed out that other technologies can decrease nations' dependence on oil, reduce pollution, relieve the burden on the electric grid, or provide opportunities for rural development. But hydrogen-based applications can provide all of these benefits. This is one characteristic of hydrogen that might differentiate it from other energy sources or technologies.

Timing of Benefits

Participants felt that it was important to discuss when the benefits from hydrogen technology could start to accrue and when investors would need to see evidence of the benefits to feel that their investments are worthwhile. As an industry representative noted at the top of the discussion, "It takes so long to get private benefits [out of a new technology]." The expected timeframe for starting to accrue benefits could help shape investment decisions, because, to the extent that the amount of the investment can influence how quickly benefits accrue, government and private-sector investors would want to ensure that potential investments are large enough to achieve the intended benefits. However, there is a difference between the timeframe that is needed to achieve benefits and the speed with which the infrastructure and technologies can be developed. The group defined a short-term timeframe as one of less than ten years and a long-term timeframe as one greater than 25 years.

Some of the participants felt that hydrogen must become a viable energy source in the short term—within ten years—for important benefits to be achieved in the medium term. These benefits, in particular, are related to air pollution and climate change, but also to the energy security benefits that could result from reducing the demand for oil. Other participants said that while it may be important for hydrogen to become viable quickly, it might need to be a mid-term undertaking, requiring ten to 25 years for full development. As a comparative timeline, participants cited the example of getting a new automobile technology to market, which takes at least ten years, and even then the technology may be introduced in a limited number of cars.

Forum participants expressed the view that short-term action is required for the following reasons:

- The opportunity for motivating a change in the energy infrastructure is here now; it may be gone in ten years.
- If long-term impact is going to be realized, short-term action is needed now.
- Benefits can grow over time, but it will be critical to address carbon dioxide issues sooner rather than later.

One idea in particular generated a good deal of discussion among forum participants—there may be market niches that exist today, such as markets for distributed generation and small-scale hydrogen production systems, that can be deployed quickly. As one analyst noted: "A small system at home would sell like hotcakes around the world. If we don't do it, someone else will [i.e., Japan, Europe, or China]. It can happen in the near term."

One participant's industry perspective was stated this way: "Market segment affects the timeframe and potential of a new technology. Some small-scale, niche applications are ready today or soon will be. Others are further away. There is a different time scale in different markets."

These market niches could provide the basis for expanding and accelerating new technology deployment. A representative of a policymaking body offered the following thought: "There is potential in the next ten years for demand for distributed, small-scale power [generation] around the world [to increase substantially] and for a couple of companies to emerge and be world class leaders. It may not have a big impact on public benefits, but companies that get a foothold can really start to shine."

As one participant observed, while the short-term impact of a new technology in terms of benefits may be small, the infrastructure would be in place for a more rapid acceleration of benefits in the future. Companies should focus on finding these niches and exploiting the opportunities they present, the participant stressed. Of course, there may be a disruptive event that changes expectations, and technologies that are in use now may not be those that are in use ten to 20 years from now.

Critical to future expectations about hydrogen technology and the analysis that may be done to assess future hydrogen energy opportunities, participants pointed out, is how fast a transition to hydrogen can happen. This transition will depend heavily on capital turnover rates (see the discussion under "Capital Cycles and Timing of Climate-Change Policy"), the mention of which led to a discussion of "adoption curves" (the timing of adoption of new technologies) and analogies to infrastructure changes. The state of an existing infrastructure and the rate of capital turnover can impact how fast emerging hydrogen technologies could penetrate worldwide energy markets.

As one participant observed, "The delivery of benefits depends on capital turnover more than it does technology. There was a compelling value proposition in locomotives. The [transition from vinyl records] to the CD was quick, though. If you can have a car with a compelling value proposition to consumers, like the Prius, even though it costs more than a similar car with a conventional engine, you'll start to see rapid turnover. Large-scale power plants are depreciated over 40 years, and a utility company will not throw out a power plant after 15 years. So, the introduction of hydrogen will depend on the amount of capital put into incumbent technologies, too."

Some participants suggested that adoption curves might be shortening. They cited examples of adoption of new technologies that happened more quickly than conventional analyses might suggest—-e.g., compact disks, the Apple iPod, and the Prius (although there was disagreement on the last item). It is possible that analogies to other products or technologies could provide some lessons for understanding how quickly hydrogen could penetrate the U.S. energy market. There were some disagreements on how quickly that might happen, as the following exchange shows:

"Look at the CD versus the LP [long-playing record]. This is arguably in the most price-sensitive segment [of personal entertainment] . . . you would have to replace a whole record collection, worth thousands of dollars sometimes. Why [did people shift to CDs]? Because there was a compelling value proposition."

"The problem with the analogy is that record companies stopped selling LPs. The latest hit wasn't on an LP anymore. Record companies accelerated the process."

Participants generally agreed that the capital turnover occurred because there was a compelling value proposition—CDs and DVDs offered superior quality to consumers, and also were easier to produce and ship than records or videotapes, which provided some value to the entertainment companies as well. A question that forum participants could not answer but that might be relevant for hydrogen is, what came first, the industry decision to make the technology shift or anticipated consumer demand? For hydrogen, the question is whether the focus should be on the specific elements that provide compelling value to consumers, or that provide business opportunities for the private sector, or both. Such questions indicate that significant analysis could be done to determine if lessons learned from these and other

Capital Cycles and Timing of Climate-Change Policy

In conjunction with the Pew Center on Global Climate Change, the RAND Corporation conducted a study that looked at the role of capital cycles—i.e., the patterns of capital investment and retirement—and their potential impact on public policy related to the changing climate. Existing capital equipment, such as electricity generation plants and transportation infrastructure, may be a significant source of greenhouse gas emissions, and much of this capital equipment is long-lasting and expensive. Some key results from the Pew Center/RAND study include the following:

- Capital has no fixed cycle, but external market conditions are the primary drivers behind a firm's decision to invest in new equipment.
- More efficient technology is not a significant driver of capital cycles in the absence of policy or market incentives.
- Investment is focused toward key corporate goals, in particular goals driven by the desire to capture new markets.
- The dynamics of capital investment and retirement can slow the adoption of promising new emission-reducing technologies.
- Policymakers may speed the pace of capital investment by pursuing polices that seem to have little immediate relationship to climate policy.

These findings are relevant to understanding the potential for infrastructure change that could lead to further deployment of hydrogen technology and the role policy can play in that regard. Many decisions on whether or not to move toward hydrogen as an alternative energy carrier will depend on how fast existing capital might turn over in order to incorporate hydrogen technology.

Reference: Lempert, Robert, Steven Popper, Susan Resetar, and Stuart Hart, *Capital Cycles and the Timing of Climate Change Policy,* Washington, D.C.: Pew Center on Global Climate Change, October 2002.

technological analogies to hydrogen, such as personal computers and cell phones, can provide lessons for both the analysis of and understanding of how quickly hydrogen can be introduced and as a guide for policymakers to understand the role that policy can play in this regard.

Concluding Thoughts

In concluding the discussion of benefits, forum participants emphasized the fact that achieving benefits will depend on public-private partnerships. No matter how soon hydrogen is needed as an alternative energy source, or how quickly it can be established within the energy sector, forum participants felt strongly that public-private partnerships will be critical for achieving the benefits they discussed. Long before benefits are realized, these partnerships are critical to research and development and to establishing the regulations, codes, standards, and infrastructure to support hydrogen. They pointed to Germany's increasing market penetration of wind-generated electricity as an example of how the public and private sectors can work together to speed the introduction of a technology (see the discussion under "Germany's Move Toward Renewable Energy").

Germany's Move Toward Renewable Energy

A recent article in *Solar Today* on Germany's renewable energy policies reported that the German government is moving toward increasing the country's use of renewable energy, perhaps by up to 50 percent by 2050. The drivers behind this development are the risks associated with

- nuclear power, which constitutes about 30 percent of Germany's current electricity generation
- climate change (Germany has ratified the Kyoto Protocol treaty on global warming; countries that sign the treaty agree to reduce their emissions of carbon dioxide and other gases)
- Air pollution
- Dependence on nondomestic sources of energy.

German policymakers see the next 15 years as "make or break" years for the transition to renewable energy and have concluded that near-term efforts to support this transition are needed. Ten years ago, Germany had no wind power; today, wind as a power source constitutes more than 6 percent of Germany's power-generation mix. This development occurred through a combination of political will, citizen involvement, scientific analysis, and an economic strategy that reflected the associated risks and allowed for market-based investment decisions.

Reference: Aitken, Donald, "Germany Launches Its Transition," *Solar Today*, March/April 2005.

One policy participant made a case for how federal and state governments are changing their approach to regulation. "Historically, we tried to advance technologies by technology-forcing regulations. This has and hasn't worked at times . . . [current] initiatives provide opportunities for all to work together. Industry now has an input into policy, unlike in the past, when it was simply regulated."

Members of the group said that if the implementation of hydrogen energy was going to happen, the applicable regulations, codes, and standards would need to be adaptable to the changing technologies and new information that will emerge over the next ten years, and only the public and private sectors working together can make this happen. It will also be important for regulations and codes, where they are needed, to regulate performance and not focus on specific technology outcomes.

Participants emphasized that both sectors, public and private, need each other as long-term stable partners, and it is vitally important for the government to be able to make long-term commitments to hydrogen if the private sector is going to make large capital investments in it. The only way to overcome the hurdles to hydrogen production and deployment (discussed in the next chapter) is for the public and private sectors to cooperate in a way they never have before.

CHAPTER THREE

Barriers to Hydrogen's Development as an Alternative Energy Carrier

The forum's third discussion session focused on the barriers that could prevent hydrogen from becoming fully developed as an alternative energy source and as a viable player in the energy markets. This discussion was from the point of view of government and private-sector investors who, due to these barriers, could be prevented from realizing all the benefits that hydrogen is capable of delivering. Understanding the potential barriers to the development of hydrogen energy can help stakeholders shape their hydrogen-related policies and investment strategies. These barriers, participants observed, are not very different from the barriers that other new and emerging technologies in the energy sector have faced and that have been overcome in reducing air pollution (see the related discussion under "Overcoming Barriers: How California Managed to Reduce Its Air Pollution"). These barriers include regulatory roadblocks, competition from other energy sources, technological and cost barriers that hinder implementation, resistance from the public, and a lack of coherent state and federal government energy policies. (This session did not include a detailed discussion of technology issues. See Appendix C for a brief discussion of technological hurdles.)

Forum participants were asked to brainstorm on key barriers that might prevent hydrogen technologies from penetrating energy markets. This chapter provides a brief summary of three key barriers that may serve to differentiate hydrogen from other energy sources or technologies in other sectors:

- Policy barriers, which include regulatory barriers and barriers to conducting quality analysis
- Corporate risk barriers, which include those related to liability and time horizons for realizing revenues from commercialization of hydrogen energy
- Public perception barriers (i.e., does the public believe energy is a problem?).

These barriers and the problems they present are independent of each other for the most part, but occasionally they interact and overlap. In fact, a fourth barrier cuts across all of the other three: the lack of a robust set of economic metrics to value hydrogen. For a full list of the barriers identified by forum participants, see Appendix B.

Overcoming Barriers: How California Managed to Reduce Its Air Pollution

While many U.S. cities continue to have air pollution problems, significant reductions in air pollution have occurred over the past few decades. (The information in this sidebar is drawn from forum discussions and common knowledge among the environmental policy community.) California has taken a leading role in developing policy solutions for reducing urban and regional pollution. This leadership role has involved a strong commitment to obtaining scientifically credible data (air quality management districts in California have some of the best air-quality data available); support for the development of leading technology (e.g., catalytic converters); a portfolio of air-quality policy options that include subsidies, mandates, incentives, and emissions trading; and markets driven by active targets and goals. The air quality in Southern California has improved over the past 20 years despite growing populations and growing numbers of cars. It is possible, therefore, to overcome the various barriers to reducing air pollution—e.g., regulatory, technological, and cost barriers—with a mix of good science, robust analysis, and technological innovation driven by sound policy.

Policy Barriers

A key policy barrier and one that makes hydrogen different from other energy technologies is the large number and variety of public organizations that may need to be involved in the development of hydrogen as an alternative energy source. In addition to energy organizations, such as the Federal Energy Regulatory Commission and state Public Utility Commissions, agencies responsible for local building and fire codes, zoning, air pollution control, and transportation also may be involved. Rationalizing the roles and responsibilities of the various federal, state, and local agencies and jurisdictions could be an important factor in the future success or failure of a transition from other energy sources to hydrogen.

Forum participants also believed that the lack of a coherent and comprehensive state and/or federal energy policy has led to confusion on the part of both the public and the private sectors over the costs, benefits, and safety issues associated with hydrogen. For example, there is no clear picture of how many refueling stations may be needed to "jump-start" hydrogen and how these stations should be structured and where they should be located. One industry representative offered the following perspective: "There is no comprehensive energy policy in the United States, nor in California. Rather, there is a series of uncoordinated steps to integrate hydrogen to bring it forward. From a global perspective, linking together these efforts, and having more continuity—a true roadmap of where we are going and why—would help in the long-term allocation of resources and long-term planning."

The lack of consistent and reasonably independent estimates of the benefits and costs of hydrogen development and implementation has hindered discussions about the viability of hydrogen investments. An industry analyst suggested that the way that hydrogen as an energy carrier is valued is not appropriate given the nature of hydrogen. For instance, he noted that the price of hydrogen is often compared with the price of gasoline: "Hydrogen has value not incorporated into the price of a gallon of gasoline, such as energy security value, [value to the] environment, etc. Another example would be the way that electricity is traditionally valued. Fuel-cell metrics were evaluated in dollars per kilowatt hour 12 to 15 years ago, based on central generators (such as a combined-cycle unit connected to a grid) as a commodity energy source using bus bar power costs of a small power plant. Today, we know that the valuation of distributed generation is very different. If you use yesterday's yardstick, you end up recreating the past."

Corporate-Risk Barriers

Conference participants generally agreed that the problem of finding short- to medium-term profitability in the energy industry poses a significant problem in attracting investment in hydrogen energy.

An energy industry consultant participating in the forum provided some background on the issue of profitability: "The good news is that there is a lot of interest in investing in clean technologies, including socially responsible investment funds and VCs [venture capital funds] that focus on clean energy. Statistics in [the energy] sector show that 5 to 6 percent of venture funds explicitly include clean technologies, half of which are energy companies (production, storage, etc.). The bad news is that profitability and production performance of clean energy businesses have not met expectations."

Another barrier is the inability of companies to earn the benefits that some technologies, such as distributed generation (DG), can provide to society. One forum participant, an energy industry analyst, noted that "the utilization of DG as a means to cope with T&D [transmission and distribution] issues is important. Demonstrable studies indicate directly what [the] benefits are. The problem is, that with plain old regulation as we know it, it is difficult for a plain old wires company to take those benefits."

The critical point for securing or maintaining investment in a new product or technology is the period between when a new technology shows progress in R&D and when it is ready as a commercially viable product. This period is difficult to finance and is often referred to as the "valley of death" (see the discussion under ""Mind the Gap: Bridging the Valley of Death"). Some participants felt that large original-equipment manufacturers (OEMs) that are currently investing in hydrogen could weather the valley of death simply because competitive pressures will force them to continue to invest in hydrogen, and they have the money to do so. Smaller or medium-sized companies would face significant problems during this period unless they receive some type of government assistance, because they likely would not be able to raise the money to continue investing in hydrogen.

Although the government could intervene at this critical point with funding, it would raise two questions: Should the government do so, and to what point must technologies progress before the government can reduce its support of their development?

As one industry representative said: "If you cannot produce anything of value to whomever you're serving in that time period [in the valley of death], it is a huge problem. You don't have to solve the end game in the valley period, but you have to produce something, depending on the length of the technology curve. It is very problematic for small entities, but better ideas may be produced by smaller entities. Public-private partnerships may help bridge this valley period."

Mind the Gap: Bridging the Valley of Death

The "valley of death" is a term that is widely used by business and policy analysts to describe the period after new-product research and development (R&D) when the product has been shown to be technologically viable but before it is proven to be commercially viable. During this period, there is a lack of funding for marketing the new product. In the initial stages of development of a product, significant opportunities exist to secure funding from the government (see the figure below). As the bulk of the research winds down, funding also declines, particularly if the government chooses not to fund demonstration and commercialization efforts. Toward the end of the R&D stage, private financing begins to pick up, including venture capital at initial stages, and then private entities take over the funding as the product moves to the commercialization stage. In essence, the valley of death is the dip in the funding continuum during which government and basic research funding declines and when private-sector investors believe the risks are still too high for large capital investments in a new product. This lack of funding during the middle stage from R&D to commercialization is believed to hinder the deployment of new and emerging technologies.

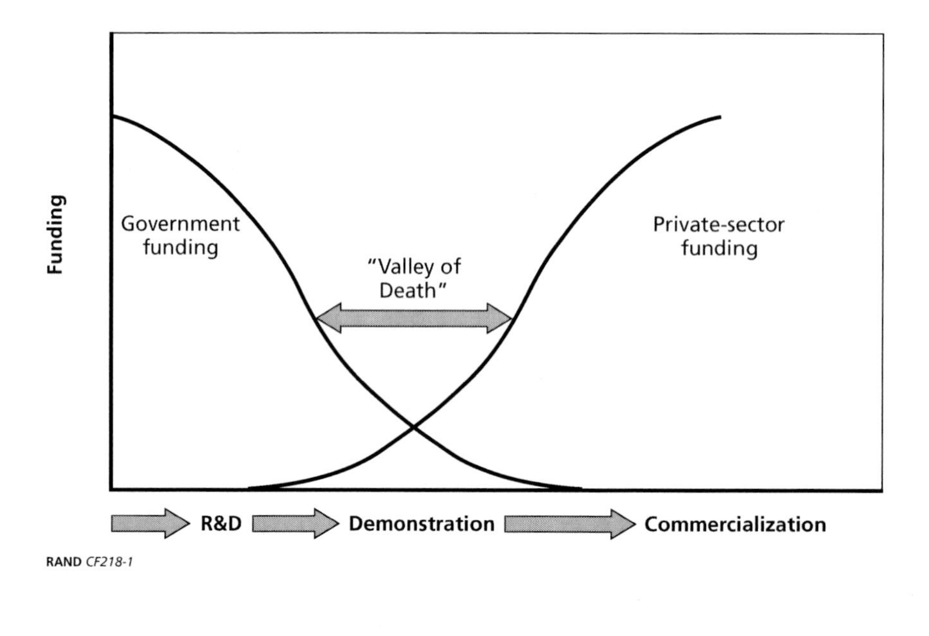

Another industry representative added: "This problem exists even for well-capitalized companies. Unless government provides investment opportunities (for example, through tax credit and investment regimes), smaller companies will continue to drop off. This is a critical area where government can intervene."

In response, a forum participant representing the policymaker perspective noted the following: "Companies must prove performance. If cost is the only issue that remains, then government can add all that up and perhaps help. However, it is probably premature to make that decision yet, regardless of whether you are considering the taxpayer or the investor."

Participants said that one contributing factor to the valley of death is the potential for significant liabilities from a new technology—i.e., some companies may be reluctant to invest in hydrogen because they fear that as the technology and infrastructure get up and running, they will not work perfectly at first, and that persons or property may be harmed, leading to potential lawsuits and other liability issues for the companies that own and/or operate the hydrogen facilities. While this is a barrier for other new and emerging technologies besides hydrogen, there may be the perception of greater liability associated with hydrogen, especially among early adopters of the technology.

In response, the argument was raised that hydrogen may not be that unique in terms of liability. A participant from the energy industry said the following: "Most industries wouldn't operate if they knew how [difficult] it was to run a utility because of regulations. There's a lot of fantasy out there. Electric operations have enormous liability. You can get sued by customers. Actions in legislatures go against utilities. There are a lot of hazards, because utilities support food [systems], life safety systems, etc." The participant also said that it would be difficult to allow those who do not want to share liability to enter the market freely. However, if there is a law that shields utilities from liability, "utilities will buy all sorts of technologies."

Participants said that some assessment of the liabilities would be useful, and those assessments could lead to policies that may limit liabilities, which could have a positive impact on corporate investments.

Another issue that was raised with regard to the "valley of death" is the need for companies to "perform," and the timeframe for demonstrating performance has been shrinking. Innovative ways for private investment to achieve some short-term returns may be necessary to bridge the investment gap.

Participants on the corporate side discussed the problem of companies lacking an understanding of the potential for the hydrogen market. The hydrogen market is more complex and perhaps more uncertain than other markets in which companies may consider making investments, and these factors can present a considerable hurdle standing in the way of corporations investing in hydrogen. Another hurdle, as one participant noted, is "the lack of understanding of the entire energy-supply chain, particularly when trying to finance a project."

For example, the natural gas industry, which could be the industry supplying the main fuel source for near-term hydrogen production, has taken little notice of and made little or no investments in researching or developing capabilities for hydrogen. It is likely that those companies do not perceive near-term opportunities for hydrogen, and they fear that investing in hydrogen diverts them from their core businesses. A more complete understanding of the opportunities that hydrogen offers the natural gas industry in terms of hedging against price fluctuations and against the potential for competition from other sources might influence some of those companies to make investments in future hydrogen technologies as a market opportunity.

Public-Perception Barriers

Forum participants discussed the fact that the public's perception of hydrogen also plays a part in whether, and how quickly, hydrogen can be developed as an alternative energy source. If the public does not understand hydrogen as an energy carrier, or perceives it to be a potential problem, it may pose a significant barrier to commercialization of hydrogen that will require a concerted effort to overcome.

The potential value of hydrogen is difficult to explain to the public, because it is not something found in the ground, and it can be produced and used in many different ways. One representative of the energy industry noted: "We have branding issues. Hydrogen means different things to different people."

In a similar vein, an industry consultant suggested the following: "The semantics used in the public debate are a barrier. For example, physicists talk about hydrogen being an energy carrier, rather than a fuel . . . [using that terminology] just obfuscates what's going on. Also, discussions of energy efficiency are in the wrong context. These discussions totally ignore why we want to do a project in the first place. For example, it is often forgotten that the efficiency of getting gasoline to your car is negative. [Meanwhile,] people are doing detailed studies of ethanol efficiency."

Some participants noted that the benefits of hydrogen (see Chapter Two) are diverse and complex, and that they are very difficult to explain to the public. The public discussion on hydrogen has sometimes obfuscated the critical issues rather than shed light on them. A policymaking representative defined part of this problem: "People aren't going to buy hydrogen because of public benefits. The key is that the fuel cell or conversion device has to offer something better [to the individual user]. The chicken or the egg [issue] is oversimplified." Another participant noted, "The technology must be better than what it's replacing, from a public perspective."

Concluding Thoughts

The discussion on barriers to hydrogen energy often referred back to issues of public perception and "message" issues—what is hydrogen and how is it used? Government, private-sector, and public acceptance of hydrogen as an energy carrier is critical to any future significant penetration of the energy markets by hydrogen. But how these groups view hydrogen is not well understood.

The lack of a robust set of economic metrics to value hydrogen is another barrier that was mentioned. The measures used to value hydrogen (and other forms of energy) are rooted in a petroleum-based economy that does not necessarily reflect the actual costs of energy. The decisions that will lead to the incorporation of hydrogen into the energy mix will likely derive from dealing with security and environmental issues, none of which are adequately valued today.

The final thoughts from both policy and industry representatives on the barriers to hydrogen focused on technology. If hydrogen is to move forward, one participant said, "You need to be technology ready, along the whole chain (from a private-sector perspective). You need to meet performance requirements, cost requirements, and reliability. . . . The big gorilla is the technical challenge, which we should circle and surround with big lights. Production from renewables, carbon sequestration, storage, and the reliable service life of fuel cells are all issues. Together, these issues multiply the risks of investing in hydrogen."

CHAPTER FOUR
Evaluating the Risks and Impacts Associated with Hydrogen-Investment Policy Options

For both government and private-sector investors, making decisions about potential investments in hydrogen requires an evaluation of the risks and impacts associated with various investment approaches and of how well those approaches might hold up in various possible future scenarios. For governmental bodies, "investment" decisions include not just those concerning how to spend public funds but also policy decisions and evaluations of which policy actions will bring about desired change. Both government *policy* portfolios and institutional or individual *investment* portfolios are intended to yield economic returns and balance risks over short-term and long-term investments. Of course, governments must respond to a wider range of problems and goals than those that exist in the private sector and ensure that the resources that will be needed to address future problems become available. Nonetheless, government policymakers can learn from the investment strategies of the private sector to assess whether significant investments in hydrogen should be part of their investment portfolios.

Despite the guidance provided by private-sector investment strategies, government policymaking is not easy. Policymakers disagree, for example, on whether and how to respond to long-term and deeply uncertain challenges that sometimes have a global reach. The stakes in such cases can be high. As the federal government, and California, look toward the future in regard to energy sources, they face substantial uncertainties, including whether there will be continued volatile and rising oil prices and the risk of short- or long-term shortages; the risks associated with increasing greenhouse gas emissions, which could include rising sea levels and reduced availability of water; and the possibility of continued air-quality problems. Further, under some circumstances, aggressive near-term actions to introduce hydrogen could have a negative impact on economic growth. Under different circumstances, these costs could be mitigated if new technologies emerge that make hydrogen much cheaper to produce and deploy than is currently possible. Conversely, failure to take near-term actions to deal with environmental and energy problems could have large economic and environmental consequences. But, these consequences could be less severe if, for example, current scientific predictions about the extent of climate change turn out to be overly pessimistic.

To assess the risks and impacts associated with various approaches to investing in hydrogen technology, forum participants were divided into three groups, each of which addressed a different broad policy approach. This exercise was limited solely to approaches that governments might take, and it was focused on California; it excluded the federal government or other states. The goal was to narrow the focus enough to have a meaningful discussion and perhaps provide some input into the extent to which California should invest in its emerging Hydrogen Highway Blueprint Plan, an initiative to support the rapid transition to a hydrogen transportation economy in California. (The next chapter of this report describes the exercise in more detail and summarizes the findings reported by the three groups.)

Exercise Format

This session's exercise, to assess the risks and impacts associated with investing in hydrogen, consisted of three interdependent elements:

- First, participants were to assume that government policymakers would take one of three approaches to hydrogen investment and hydrogen-related policymaking—a laissez-faire approach that would rely on market forces to make hydrogen a viable part of the energy market, a very aggressive approach in which policymakers would pursue opportunities and take risks to accelerate hydrogen's establishment in the energy market, and another approach that fell in between the two.
- Second, the exercise utilized four possible future scenarios. The first was a "no problem" scenario in which there are no significant energy, environmental, or economic problems. In the second scenario, environmental problems predominate, but energy problems are minor. In the third scenario, the opposite is true—energy problems predominate, but environmental problems are minor. In the fourth scenario, called the "big problem" scenario, all of the above are a problem—the environment, energy, and the economy.
- Finally, the exercise outlined some presumed goals of the state's policymaking approaches regarding hydrogen, beyond just making hydrogen a part of the state's energy profile. We posed to the group that the government has four main goals that they would want to achieve with any hydrogen policy, and the group was asked to assess the risks of their policy as they relate to these goals. The four goals were: to improve energy security, to reduce the impact of climate change, to reduce air pollution, and to improve economic growth.

It should be noted that, given the obvious time constraints, this exercise was not comprehensive—i.e., participants could not address all the possibilities and variables that would apply in the decisionmaking process. For example, in reality, the approaches governments have to choose from would not fall neatly into three categories; nor would governments have to choose just one approach and stick with it. Many approaches, and a variety of specific actions—in some cases, even actions unrelated to hydrogen—might be effective in helping government and private-sector investors achieve their hydrogen-related goals. This exercise was intended to demonstrate that a method exists for assessing the impacts and risks of various approaches, while acknowledging that there are considerable uncertainties in any approach.

The Three Approaches to Hydrogen Investment and Policymaking

For this exercise, we proposed the following three general approaches to government investment and policymaking in California. None of these scenarios is based on modeling or forecasting analyses, and none of them is meant to suggest specific predictions. Rather, they describe a broad range of possible outcomes.

- **Market-only.** In the market-only approach, the government would take no action to make hydrogen a viable part of the energy market. For example, it would step away from funding hydrogen demonstration and deployment projects. Hydrogen would not penetrate the energy markets significantly before 2050.
- **Moderate action.** This approach could by 2020 result in
  - 150,000 hydrogen-fueled vehicles on the road in California
  - 5 percent of electricity demand in California fueled by hydrogen
  - 50 percent of hydrogen produced from coal or nuclear sources.
- **Aggressive action.** This approach could by 2020 result in
  - one million hydrogen-fueled vehicles on the road in California
  - 20 percent of electricity demand in California fueled by hydrogen
  - All of the hydrogen produced would be climate neutral, as compared with alternatives that would not be so, and half of the hydrogen would be produced by renewable resources.

Future Scenarios

The impacts and risks of the three approaches above and the actions they imply depend on what the future holds for the energy supply, energy prices and their impact on the economy, and environmental concerns such as climate change and regional air pollution. Because it is impossible to forecast the future with any reasonable accuracy, we suggested, for discussion purposes, four different "futures" that California might find itself in 15 years from now. Each of the three policy approaches would have different risks and impacts depending on what the future holds. The four future scenarios are as follows:

- **No problem.** By 2020, climate impacts will be mild, regional air quality improves, energy prices are stable, and supplies are adequate.
- **Environmental problem.** By 2020, scientific studies are more convincing that climate impacts will become severe, and regional air quality continues to deteriorate, but energy prices are stable and energy supplies are adequate.
- **Energy problem.** By 2020, climate impacts will be mild, and regional air pollution improves, but energy prices are highly volatile and energy supplies are disrupted.
- **Big problem.** By 2020, there are both environmental and energy problems. Scientific studies are more convincing that climate impacts will become severe, urban and regional air quality continues to deteriorate, and, at the same time, energy prices are highly volatile and energy supplies are disrupted.

In these scenarios, if the government were to take aggressive action immediately, the impacts and risks would play out differently with a "big-problem" future than they would with a "no-problem" future. If the government were to take aggressive action in a big-problem future, then it would have already taken measures to reduce emissions, local impacts would be less, and oil consumption would be reduced, which means that volatility in prices would have a smaller impact. On the other hand, in a no-problem future, there are likely to be some investments in technologies that are not used or are not cost-effective, and investments made in hydrogen would have less of a payoff than investments made elsewhere.

Goals for the California Government's Hydrogen Investment and Policymaking

Governments generally have goals for their policy actions. In considering whether to make significant investments in hydrogen, governments should look at how these investments impact critical outcomes. RAND defined some measurable goals for the participants to consider:

- **Improving energy security** as measured by reduced energy price volatility (as a proxy for a broader definition of energy security)
- **Reducing impacts of climate change** as measured by net change in emissions of greenhouse gases (as a proxy for impact of climate change)
- **Reducing air pollution** as measured by net change in urban air pollution
- **Improving economic growth** as measured by net change in gross domestic product.

Each group was asked to consider these outcome measures when they assessed the impacts and risks of their assigned policy approach in the context of each of the four future scenarios. For example, if the California government chooses a market-only approach and the state ends up with the big-problem scenario, and if other states or other countries besides the United States have made investments in hydrogen or other alternative energy sources, then California could lose a comparative advantage in energy technology and face higher energy costs than those of other states or foreign countries. There would also be the risk of higher energy costs due to oil price volatility and health risks from pollution-related problems. On the other hand, if California ends up with the no-problem scenario, it would not have made investments that perhaps others had made.

Findings from the Exercise

As stated above, forum participants were divided into three groups, and each group was assigned one of the three policy approaches for consideration. The purpose of the exercise was to compare the approaches and determine if one of them is more "robust" for some scenarios and some outcomes than others. A robust solution is one that has few if any downside risks or negative impacts as it applies to all scenarios and outcomes. Approaches that have some serious downsides are not as robust as those that minimize the negative outcomes.

Each group was asked to color-code one of three matrices to indicate their opinion of the impacts from the market-only, moderate, or aggressive policy approach. (The matrices, which were created by RAND for the forum, are presented as grayscale versions in Appendix E.) The matrices in the appendix display the level of impact for various investment and policy goals given the four future scenarios. The color black indicates a large and negative impact as a result of a particular approach, the color gray indicates that the impact is negligible one way or another, and the color white indicates a large and positive impact. In some cases, the group did not agree on the potential outcomes or they decided that they needed some in-between categories, so some of the boxes include two or three colors. For example, a square that is white on the top and gray on the bottom indicates that there is a chance of positive impacts, but that they are not likely to be large. If one color predominates in a box, but there is another color in its bottom-right corner (as in the box in the lower-right corner of Figure

E.1), the group felt that one outcome (indicated by the dominant color) was most likely, but there is a small chance that another outcome is possible.

Impacts of a Market-Only Policy Approach

For the approach in which the government funds only R&D and allows market forces to run their course, the group was of the opinion that there would be significant risks to the environment and to the economy with the big-problem and environmental-problem (climate change and regional air pollution) scenarios.

A forum participant representing the policymaker perspective noted the following: "The big risk is that you don't know that you will have a problem in 2020 until you reach 2020. You're responding to it in short-term market solutions. There will be more volatility created by dealing with the supply curve marginally." An industry representative elaborated on that point: "It depends on how problems manifest themselves. It depends on whether the problems manifest themselves incrementally (in which case the market is more efficient) or whether there is a huge market disruption (then the market will not be capable of reacting fast enough, or doesn't effectively address issues)."

The group believed that under the energy-problem scenario, market forces would respond quickly enough to generate some positive impacts. The group envisioned some short-term economic disruptions; therefore, a portion of the economic-growth matrix is red (see Figure E.1). The group thought that there would be some potential positive impacts in the no-problem scenario, primarily driven by outcomes from R&D that could be applied to other areas, but mostly the impact would be neutral.

Impacts of a Moderate Policy Approach

The group that discussed the moderate policy approach disagreed about the potential impacts of moderate action on the part of the government. For the scenarios other than the big-problem scenario, the group saw some positive impacts on one measure—economic growth—as a result of moderate action, but not much in the way of impacts on the other measures. On the big-problem scenario, however, there was significant disagreement. Some in the group said that moderate actions would be enough and that, as big problems hit, the state would be ready to address them quickly and efficiently, and consequently, there would be positive impacts. Other members of the group said that these actions would not go far enough to prepare the state for the big problems and would not create enough infrastructure to achieve positive benefits, and that the impacts of the problems would be negative.

The following is an exchange between industry and policy representatives discussing this issue:

"The ramp-up time will be shorter [than previously], but otherwise you're rearranging the chairs on the Titanic. You will have a learned-by-doing experience, you'll have addressed the regulatory issues, and [you] will have an impact by demonstrating technology."

"You remove the 'first provider' hesitancy, but it's a band-aid on a hemorrhage."

"I'm not sure why it's a band-aid. This implies you are ignoring the problem, but you've positioned yourself for an upswing, and since there is so much unknown, you at least have to throw resources at it and see what's happened."

Although two members of the group remained neutral, the debate was both informative and spirited and could have gone on longer than the forum's schedule allowed.

Impacts of an Aggressive Policy Approach

The group that addressed an aggressive policy approach saw large positive benefits in the big-problem scenario and some positive benefits in the other scenarios, with potentially some small negative impacts to the economy in the environmental-problem scenario. In the no-problem scenario, an aggressive approach makes the energy system even more efficient and reduces environmental problems even further, so there is still some potential for positive impacts. However, the group felt that there would be significant risks to the state's economy if energy costs were higher in California than in other states or in foreign countries, and significant risks of opportunity costs associated with investments that were not needed (i.e., money for those investments might have been better spent elsewhere).

The group recognized that developing policies to achieve positive outcomes would not be easy, and they spent considerable time discussing the types of policies—from taxes on carbon emissions, to clean-air credits, to education in public schools—that could be put in place to achieve certain benefits. One industry representative added, "[It is] important to look back to the past at what didn't work and make sure we don't repeat the same mistakes."

Concluding Thoughts

In examining the figures in Appendix E, some observations can be made. If there is a chance of having environmental problems, then the market-only approach is never a robust strategy for the government to follow, because there are significant downside risks (see the accompanying discussion under "Making Policies Robust"). These risks include not only the direct

Making Policies Robust

In a recent *Scientific American* article, Popper, Lempert, and Bankes (2005) discuss a new approach to developing robust long-term planning. The authors posit that a robust planning strategy performs well when compared with alternative strategies across a wide range of plausible futures. A robust strategy need not be the optimal strategy in any future scenario; it will, however, yield satisfactory outcomes in both easy-to-envision futures and difficult-to-anticipate contingencies. This approach replicates the way people often reason through complicated and uncertain decisions in everyday life. They seldom plan for an optimal outcome. Rather, they seek strategies that will work "well enough" to hedge against various potential negative outcomes.

As the authors of the article point out, incorporating robustness into decisionmaking was not previously possible because it would make the decisionmaking too complex. Today, through a combination of human interaction and modern computing capabilities, robust long-term planning is a possibility.

Reference: Popper, Steven W., Robert J. Lempert, and Steven C. Bankes, "Shaping the Future," *Scientific American,* April 2005.

risk of environmental problems, but also risks associated with losing technological advantages if other states or other countries developed technologies to deal with these problems.

If the no-problem scenario is unlikely, then the aggressive policy approach would be robust, with small downside risks to the economy given the environmental-problem scenario.

The moderate policy approach could be the most robust if one believes that the moderate policy actions would move California and the nation far enough toward the direction of increased use of hydrogen to alleviate potential problems should the big-problem scenario emerge. However, if one believes that those actions will not be enough to avoid big problems, then this scenario is not a robust solution.

CHAPTER FIVE

Information Needed for Decisionmaking by Public-Sector and Private-Sector Investors

As was stated at the top of this report, the purpose of the RAND Forum on Hydrogen Technology was to engage experts with an interest in hydrogen as an alternative energy source in an open discussion on the subject and to identify analyses and actions that the public and private sectors need to inform their decisionmaking about hydrogen. The forum ended with a discussion designed to integrate the findings and feedback from the earlier sessions and extract a set of issues and recommendations for engaging decisionmakers in a discussion on hydrogen and guiding future analyses. This chapter provides a sample of the comments from participants during the wrap-up session, and it describes a recommended set of analyses and actions that constitute next steps in an effort to evaluate hydrogen as an alternative energy source.

Sample Comments

"Today, an entrepreneur, a lab, or a big company that decides to embark on commercialization of a device, subsystem, or control software faces a lack of economic references, and therefore, what they encounter is that every person in the process is padding his own estimates, and the aggregates of that padding debilitate progress," said an industry consultant.

Some sense of certainty about government commitment is necessary before funding bodies will make substantial investments in that technology. This is true not only for companies in the private sector with a stake in hydrogen, but also for regulators, who need to have consistency in their actions and a long-term perspective on hydrogen. One industry participant said, "If I'm going to throw you the football, are you ready to catch it, from a societal and customer-acceptance point of view?"

Continuing on that theme, an industry representative added, "You need to know you can make money," to which another industry representative replied, "Eventually." And another said, "While I don't know the future, the question is, what is the ideal portfolio, and where does hydrogen fit into my portfolio? The investment is like the share of bonds in my retirement fund."

Banks have a role in loaning money to fund investments in hydrogen, an industry analyst pointed out. "The acid test is the bank. Banks will ask you to show them the warranties, performance guarantees from OEM, liquidated damages, and that you have this [venture] insured. The bank may then consider you, if you have these four criteria. Even then, other investments might have higher ROI [return on investment]. The thing that we forgot in the gas turbine arena, for example, was that people went along with the technology because GE and Westinghouse were behind it. [But] the insurance industry had to pay

$450 million in claims. Now [banks] require liquidated damages and longer warranties. If you are selling a new technology, you have to be able to convince not only the purchaser, but also the financier."

From the public sector standpoint, participants noted, disruptive forces, such as climate change, are the wildcards that can change everything. Therefore, a greater understanding of the impacts of hydrogen adoption could generate a clearer picture of the role hydrogen would play in the states' energy portfolios. But as one policy representative said, "Barring a disruptive force, hydrogen needs to be better than the previous infrastructure. We need to know whether the government can change the paradigm [of energy supply choices], short of a disruptive force."

Implications for Public-Policy Decisionmakers

From a public policy point of view, a consistent message throughout the forum was that using hydrogen as an energy carrier could provide substantial benefits for California and for the United States as a whole. Participants felt that there are substantial risks associated with not taking near-term actions toward hastening the introduction of hydrogen as an energy carrier, and the benefits could be significant. However, participants said that more information is needed to help them decide what role, if any, the government should play in furthering the development of hydrogen. The U.S. Department of Energy's Hydrogen Posture Plan and the California Hydrogen Highway Blueprint Plan are both good jumping-off points, but participants said that more objective, and more transparent, information needs to be made available to public- and private-sector decisionmakers, and that the government should assist in the development of this information. Some suggestions from participants along these lines include the following:

Create Continuity in Government Policy. Industry representatives at the forum said that the biggest problem associated with decisionmaking on hydrogen is that there is no continuity or clarity in government energy policy and no clear signal that the government intends to move forward with the development of hydrogen as an energy alternative. One of the participants drew an analogy to football: Companies not only need to know if someone will be there to catch the ball, they need to be certain that someone knows the pass pattern—i.e., that if they invest in hydrogen, they need to know that state and federal governments will partner with them to develop an infrastructure for its use, implement policies that support its use, and help plan for a future that includes hydrogen. This kind of partnership and planning will help companies determine if there is money to be made (hopefully sooner rather than later) in hydrogen technology. Participants thought that a clearer set of pathways showing how it is possible to "get from here to there" would be helpful.

Assess Current Technology Readiness. The group agreed that there is a clear need for a better assessment of technology readiness. Technological readiness is fundamental to near-term adoption of hydrogen. While the participants were bullish on the technology prospects for hydrogen, there was still considerable disagreement among the group about some of the hydrogen technologies in terms of what stage of commercial readiness they have achieved. Recent studies and technology roadmaps developed by the U.S. Department of Energy have helped to answer questions about technology readiness, but participants added that those roadmaps need to be updated and expanded to be more inclusive. Comparisons with successful technology developments of the recent past could be used to help guide the technology roadmap effort.

Understand Public Perceptions. Participants also said that it was important for policymakers to gain a better understanding of public perceptions of energy and environmental problems and the role hydrogen can play in the energy sector. Questions for which the participants did not have an answer are whether the public perceives that there are energy problems that need to be dealt with now and whether the public believes that environmental problems are severe enough that the government should take action to promote hydrogen as a cleaner fuel source.

Inform the Public. The government has a responsibility to inform the public of issues that impact their well-being and to provide objective information. There was a general feeling among participants that the government should devote resources to better inform the public about both future risks and opportunities and the options that exist to mitigate the risks associated with energy. Participants also believed that the government has a role to play in helping to clear up public misconceptions about the safety of hydrogen and inform the public of the benefits that hydrogen could produce.

Implications for Private-Sector Decisionmakers

The private sector could use the information developed in studies such as the technology roadmap studies mentioned above to determine the viability of investing in and the long-term prospects for hydrogen. In addition, participants said that private-sector investments often do not reflect the risks associated with current patterns of energy use, and that these risks should be incorporated into decisionmaking on hydrogen. Participants noted that it was important that companies better understand the liabilities associated with pollution generated from conventional fuel sources and the risks of depending on a single fuel source (i.e., volatile prices and supply problems). Participants also mentioned that it was important to engage the insurance industry in the future development of hydrogen—for example, so that the insurance industry can hedge against environmental liabilities or impacts from potential disruptions in conventional supplies.

Participants said that it is critically important for companies that are already engaged in the development of hydrogen technologies to demonstrate that these technologies are reliable and that the companies have the ability to warranty their "product," thereby reassuring the financial community that hydrogen is safe, reliable, and viable. The demonstration of this capability on the part of industry will be critical to attracting support from both public-sector and private-sector investors. There may be a possible partnership role that government can play with companies to help insure good outcomes for investors at the early stages of the adoption of the new technologies, but the primary responsibility for demonstrating reliability falls on the private sector.

A specific idea for analysis was voiced by a forum participant with a policy and industry perspective: "It might be helpful to a number of hydrogen interests and producers of technology products to see if we could do a definitive study on four or five specific stationary applications of hydrogen fuel cells that have a chance to compete cost-wise, and discover in what circumstances they could compete, so that people looking to market products in the near term have a better shot at meeting their targets."

Implications for Both Public-Sector and Private-Sector Decisionmakers

While, ideally, all the suggestions that forum participants put forth would work better if they were carried out through a public- and private-sector partnership, two key recommendations would definitely require public-private partnerships and cooperation among diverse groups with varying viewpoints–(1) shifting the analytical paradigm and (2) conducting independent and transparent analysis to answer the many questions that arise about hydrogen.

Analytical Paradigm Shift. Participants mentioned that the framework for analysis and the framework for policy and corporate investments may need a "paradigm shift." It is possible that hydrogen as an energy source will not succeed if the innovation path is based on previous paths associated with energy technology development. Alternative fuels have largely failed to gain an appreciable market share, and new technologies have had a long and slow development and commercialization process. Can public-private partnerships change the paradigm and show how the transition to hydrogen can be more like the relatively rapid transition to personal computers and cell phones? Participants said that it was important to direct the nature of the analysis and debate away from the conventional petroleum-centric view to one that reflects a broader set of costs, benefits, risks, and rewards.

Independent and Transparent Analysis. As one industry analyst said, "[Much of] the existing work has not been done by honest brokers, but by people who have something in particular they want." The group was in unanimous agreement that a rigorous, objective, and independent valuation of the lifecycle costs and benefits of hydrogen as compared with other alternative fuels and incumbent technologies was needed. Further, the analysis needs to have "open" access—i.e., transparent models and analysis that can be evaluated and replicated. The analysis needs to take into account differing viewpoints and evaluate the consequences of a variety of policy and investment actions, assessed against a number of future scenarios. The group's view was that while this analysis should build upon previous analyses, previous attempts have been incomplete or potentially biased, and, more often than not, were not open and replicable.

Concluding Thoughts

There seemed to be general agreement among forum participants that sooner is better than later for the public and private sectors to get serious about investing in hydrogen as an energy carrier. While there were differing opinions on how large today's hydrogen energy market would be, the general opinion was that sufficient technological improvements have been made in the past few years to make the hydrogen energy marketplace viable for commercial development. However, the development of hydrogen energy needs a boost from government policy, and policymakers still need convincing to move aggressively forward. They need to see more clearly the unique potential benefits of hydrogen, the new opportunities for investments and jobs, and how a portfolio of policies and investment options can meet short-term and long-term goals. Critical risks and liabilities stem from California's and the nation's dependence on a single energy source for transportation needs, from climate change and local air pollution, and from potentially reduced reliability of the electricity supply. While hydrogen as an energy carrier is not the only technology and market opportunity available to investors, participants said that hydrogen nevertheless should be a significant part of the U.S. public and private investment portfolio.

APPENDIX A

Background Information on Hydrogen

This appendix provides general background information on hydrogen—what it is, how it is produced, and what its current and potential applications are. This appendix also describes some technological hurdles to the use of hydrogen as an energy source.

What Is Hydrogen?

Hydrogen has the number-one spot in the periodic table of elements. It is the most abundant of all the chemical elements in the universe. Although pure hydrogen is a gas, very little of it is found in the atmosphere. On earth, most hydrogen is found in combination with oxygen in the form of water (H_2O), but it is also present in almost all organic matter, such as living plants and energy sources such as petroleum and coal.

How Is Hydrogen Produced?

Hydrogen (H_2) is often described as an energy source, but it is more accurately defined as a "refined fuel" or an "energy carrier." Hydrogen is not a primary energy source, in the sense that it is not found readily in nature and cannot be physically mined or extracted from geological formations. Rather, hydrogen must be obtained through a transformation of molecules or "produced" by a series of controlled chemical or biological processes that involve significant inputs of both energy and hydrogen-rich molecules.

Currently, hydrogen is almost exclusively produced from natural gas, although heavier fossil fuels and water can also be used for this purpose. Natural gas is considered to be the most favorable fossil-fuel feedstock for hydrogen production due to its high hydrogen-to-carbon ratio, widespread supply infrastructure, and ease of use. Producing hydrogen from natural gas typically involves a high-pressure, high-temperature reaction in the presence of steam and a nickel catalyst (a process known as *reforming*), but hydrogen can also be produced with oxygen (*partial oxidation*), or through a combination of both (*autothermal reforming*). Hydrogen is also a by-product of several petrochemical manufacturing processes and is produced to a much lesser extent from coal gasification, partial oxidation of petroleum, and *electrolysis* (the process of separating hydrogen from oxygen in water).

Hydrogen can be produced, in theory, from a variety of sources including primary energy sources and water. In the absence of significant technological breakthroughs in renewable electricity production or unconventional hydrogen-production techniques, natural gas and other fossil fuels will likely continue to be used to create the vast majority of hydrogen in the next decade. However, a major shift toward the use of hydrogen with the emergence of a robust fuel-cell vehicle market may pose challenges to the natural gas industry's ability to accommodate additional demand for natural gas for hydrogen production. If natural gas prices rise due to increased demand, coal-based or nuclear-based options may emerge as viable substitutes for the production of hydrogen in the near to medium term.

Although hydrogen production technologies currently exhibit significant economies of scale, the demand for hydrogen as an energy carrier would occur in new, relatively small, geographically dispersed markets. Thus, distributed hydrogen technology applications could emerge to address nascent markets outside of the traditional markets for petrochemical and refinery applications. Distributed hydrogen could be produced primarily through natural gas reforming and electrolysis in regions of the country where it is economically favorable. The hydrogen produced in such a way would have a higher unit cost, but would be a much less risky investment. Thus, the initial hydrogen supply chain would be highly regionally heterogeneous (how it is produced and moved would differ regionally) and would depend on local energy infrastructure endowments, energy commodity prices, and regulations.

Uses of Hydrogen

Hydrogen is now used primarily to produce ammonia and methanol, and to upgrade and desulfurize petroleum products at refineries. Hydrogen is also used in the manufacture of semiconductors, in food processing, and in the production of ammonia-based fertilizers.

Hydrogen may be used in a number of energy-related applications—in stationary power generation, as a blend with natural gas for low–nitrous oxide (NOx) applications, as an energy storage mechanism in regions where peak-shaving (reduction in the peak demand for electricity) is important or where remote wind or solar power (located at a distance from the source of demand) is prevalent, or for hydrogen vehicle refueling.

One transportation-sector application for hydrogen energy is in fuel-cell vehicles. Currently, transportation fuels are derived almost entirely from crude oil, and hydrogen may provide an opportunity to diversify transportation fuels. Fuel cells are highly efficient electrochemical energy-conversion devices that consume hydrogen and oxygen to create electricity and heat, with steam as the sole emission. Another application for hydrogen is stationary power generation (as opposed to power generation for a moving vehicle), particularly smaller-scale, distributed applications. Fuel cells can be used in small-scale power-generation applications, perhaps located near power-demand centers. Although some current experimental fuel cells for stationary power generation are able to operate directly from methanol or natural gas (and therefore with some carbon-containing emissions), it is expected that fuel cells for cars would need to be smaller than those for stationary power generation, and they would not be able to use methanol or natural gas directly.

Hydrogen can also be used in modified internal combustion engines, turbines, and residential natural gas burners. For example, BMW has introduced a research-scale, internal-combustion engine vehicle that can run on pure hydrogen. Turbine and residential applications, if they emerge, would most likely use a mixture of natural gas and hydrogen. Under certain conditions, the addition of small amounts of hydrogen to natural gas can lessen NOx emissions during combustion.

Future potential markets for fuel-cell vehicle refueling and distributed-power generation (small-scale generation located close to where there is demand for power) will require an infrastructure for the production, storage, and transport of hydrogen. Such an infrastructure might be geographically heterogeneous, depending on existing energy supply chains, local energy prices, and local regulations. Large-scale, centralized hydrogen production facilities could be located in remote areas near fossil fuel, nuclear, biomass (organic matter), or renewable resources, and potentially near geological formations that allow for carbon sequestration from fuel if fossil fuels are used for making hydrogen. The hydrogen that is produced could be stored as a compressed gas at several hundred times the normal atmospheric pressure or as a cryogenic liquid. Large quantities of hydrogen could be transported through pipelines. Smaller quantities could be transported by tube trailer truck or in liquid form by rail or truck. Potential breakthroughs in solid-state storage of hydrogen may favor truck or rail transportation over pipelines in some cases. Alternatively, hydrogen can be produced locally at the site of vehicle refueling stations or "energy stations," or even in homes through small-scale natural gas reforming or electrolysis. Distributed generation of hydrogen could avoid the need for a transportation infrastructure, but would still require storage and dispensing equipment. As the market for hydrogen develops, networks of hydrogen refueling stations might emerge. For example, some stations might produce excess hydrogen and ship that hydrogen to stations with storage capability, rather than production capability, producing a "hub-and-spoke" hydrogen supply network.

Who Produces Hydrogen?

The United States produces more than 50 percent of the 220 billion cubic meters of hydrogen produced worldwide each year. World hydrogen production doubles approximately every decade, mostly due to increasing demand for hydrogen by oil refineries; demand growth is stagnant in other industries.

Because existing hydrogen technologies exhibit significant economies of scale and high transportation costs, most hydrogen is produced at large centralized facilities and is consumed on site or in proximity to existing hydrogen pipeline networks near the U.S. Gulf Coast.

What Are the Major Technological Hurdles?

The technological hurdles in the development of hydrogen as an alternative energy source are related mostly to the costs of and practical barriers to building adequate infrastructure for production and storage of hydrogen. The high cost of fuel cells is another hurdle.

The need to produce hydrogen efficiently and cost effectively is a key factor in the development of a market for hydrogen. Hydrogen production from fossil fuels is a mature industry. Hydrogen production has been achieved cost effectively because it is done on a small scale and at the place where the hydrogen is needed, so there is no need for distribution. Ramping up hydrogen production cost-effectively to compete with other energy sources may prove to be difficult.

Distribution and dispensing of hydrogen, and related public safety concerns, are other key infrastructure challenges. The petrochemical industry has experience with hydrogen pipelines and tube trailers, including a pipeline network near the Gulf of Mexico in support of the refining and petrochemical complexes in the region. However, the construction of additional pipelines near densely populated areas poses safety issues. In particular, the retrofitting of natural gas pipelines for transporting hydrogen or blends of hydrogen and natural gas raises safety issues associated with leaking of hydrogen.

Perhaps the greatest hurdle in the development of a hydrogen infrastructure is hydrogen storage that would allow smaller, lighter fuel tanks and more efficient transport. Currently, hydrogen is stored as a compressed gas at 800 times the normal atmospheric pressure or as a cryogenic liquid that takes up considerable storage space and consequently is expensive to transport. Hydrides and more exotic solid-phase storage technologies utilizing carbon nanotubes are in the early research phase and therefore are highly speculative.

As stated above, another major hurdle in the development of a hydrogen market is the high capital cost of fuel cells. Although part of the unit cost of a fuel cell can be attributed to the small number of fuel cells that are available, there are nevertheless major technological hurdles that would make them more expensive than conventional energy-system alternatives even if they were mass-produced. Currently, research is being conducted on several different types of fuel cell systems, each of which has different characteristics and would be appropriate for specific electric and thermal load profiles. One of the issues with the polymer electrolyte membrane (PEM) fuel cell, the leading contender for vehicle applications, is the amount of platinum catalyst it requires. Companies such as General Motors are pursuing ways to integrate fuel cell technology into vehicles that are stylish and equipped with the latest on-board electronics as well as being fuel-efficient. Other end-use applications of hydrogen, such as internal combustion engines and turbines, have received less attention than fuel cells but may prove to be important markets for hydrogen, particularly if they can be achieved at much less cost than fuel cell technologies.

References

California Hydrogen Highway Web site, State of California (http://hydrogen highway.ca.gov/).

Hydrogen, Fuel Cells & Infrastructure Technology Program Web site, U.S. Department of Energy (http://www.eere.energy.gov/hydrogenandfuelcells/).

The Hydrogen Economy: Opportunities, Costs, Barriers, and R&D Needs, National Research Council and National Academy of Engineering: Washington, D.C., 2004.

Lipman, T. E., *What Will Power the Hydrogen Economy? Present and Future Sources of Hydrogen Energy,* Final Report, prepared for The Natural Resources Defense Council, Davis, Calif.: Energy and Resources Group and Institute of Transportation Studies, University of California, Berkeley, and Institute of Transportation Studies, University of California, Davis, UCD-ITS-RR-04-10, July 12, 2004.

Lipman, T. E., G. Nemet, and D. M. Kammen, *A Review of Advanced Power Technology Programs in the United States and Abroad Including Linked Transportation and Stationary Sector Developments,* Final Report, prepared for the California Air Resources Board and the California Stationary Fuel Cell Collaborative, Energy and Resources Group, University of California, Berkeley 2004.

Rifkin, J., *The Hydrogen Economy: The Creation of the World-Wide Energy Web and the Redistribution of Power on Earth*, New York: Jeremy P. Tarcher/Putnam, 2002.

Romm, J. J., *The Hype About Hydrogen: Fact and Fiction in the Race to Save the Climate*, Washington, D.C., Island Press, 2004.

APPENDIX B

Perceived Benefits from and Barriers to Using Hydrogen as an Alternative Energy Source

Tables B. 1 and B.2 list the perceived benefits from and the barriers to using hydrogen as an alternative energy source, which were discussed in Chapter Two and Chapter Three, respectively.

Table B.1
Perceived Benefits of Hydrogen Cited by Forum Participants

General Category of Benefits	Examples	Other-Technology Opportunities
Reduced oil consumption	Diversify transport fuels Reduce trade deficit Reduce international tension Reduce risk of water contamination Reduce air pollution Extend life of natural resources Reduce waste, including toxic waste Create potential for more-predictable costs	Compressed natural gas, biofuels, hybrid vehicles, fuel-efficient vehicles, and ultra-low-emission vehicles
Electricity generation	Improve transmission and distribution efficiency Defer transmission and distribution investments Provide storable electricity Reduce contingent power costs Complement electricity as a carrier Create opportunities for remote power	Microturbines, photovoltaics, and wind generators Compressed air Small-scale technologies
Environmental benefits beyond reduced oil consumption	Reduce greenhouse gas emissions	Renewables
Social benefits for developing countries	Create a sense of optimism about the future Retain local wealth Create potential for spin-offs Improve competitiveness with other countries Incur lower societal costs than would be incurred with other energy forms Revitalize other sectors, such as agriculture (e.g., using biomass to produce hydrogen) Create potential new products and markets Increase safety	

Table B.1—Continued

General Category of Benefits	Examples	Other-Technology Opportunities
Other benefits to developing countries	Leapfrog to cleaner technologies	Renewables and small-scale electricity generation
	Provide efficient rural energy and water services	
	Reduce urbanization pressures by providing rural opportunities	
	Provide multiple uses and means of production	
	Create multiple storage opportunities	
	Provide multiple-scale production opportunities	
	Offer benefits to indigenous populations	
Private-sector benefits	Provide more-secure and higher-quality electricity	Back-up generators
	Become more profitable	
	Help companies meet environmental and other regulations	
	Reduce electricity costs—in particular, demand charges	
	Reduce price uncertainty	
	Increase safety	

Table B.2
Perceived Barriers to Using Hydrogen Cited by Forum Participants

General Barrier Category	Examples
Technology	Advancement of competing technologies
	Lack of infrastructure
Policy	Lack of national policy and political will
	Competing social interests
	No coherency in standards and codes
	Institutional barriers to acceptance of new technologies
	Length of time to realize the public benefits
	Competing subsidies and lack of a level playing field
	Differing measures of success in public and private sectors
Costs and financing	Technology challenges and cost barriers
	Lack of large capital financing
	Possibility of stranded assets
	Perception of a zero-sum "energy game"
	Risks and liabilities faced by utilities
Public perception	Public perception of risks
	Branding problem (public understanding of what hydrogen is and what it can do)
	Public perception of level of safety
	Lack of public understanding of the actual cost and efficiency of gasoline
	Question of whether consumers perceive hydrogen to be new and valuable
	Lack of appreciation of the value of hydrogen
	Lots of myths, not lots of facts
	General lack of acceptance that fossil fuels are limited energy sources
	Confusing semantics used in public debates on hydrogen

APPENDIX C

Forum Agenda

8:00 a.m. **Breakfast and check-in**

8:30 a.m. **Welcome, statement of purpose, and introductions**

8:45 a.m.–9:30 a.m. **Setting the stage—Various perspectives on the benefits and costs of hydrogen**

- Industry
- Policy
- Business investment
- Technology
- Valuation

9:30 a.m.–10:30 a.m. **Why hydrogen? (i.e., What are the potential benefits?)—Differing perspectives and the rationale for hydrogen**

- Public sector or private sector
- Short term or long term
- National, local, or international

10:45 a.m.–11:45 a.m. **What are the obstacles to introduction of hydrogen?**

- Is there a "valley of death?"
- Are technologies ready for prime time?
- How long is too long for profitability in the private sector?
- Can the public and private sectors really work together?
- Is hydrogen development too fractured?

11:45 a.m.–12:00 p.m. **Breakout session: Four Scenarios for Assessing Future Risk**

Breakout groups will use four future scenarios regarding energy and the environment to assess the risks of various options relating to energy security, climate change, local air pollution, and economic growth.

12:00 p.m. **Lunch**

1:00 p.m.–2:30 p.m. **Three breakout sessions**

Members of each group will assess the risks of various policy options and their and impact on California and the United States as a whole.

2:45 p.m.–3:30 p.m. **Five-minute reports from each team, followed by a discussion period**

3:30 p.m.–4:15 p.m. **What do you need to know to make a case for (or against) a near-term, more rapid investment in hydrogen?**

- What measures would you use?
- What do you know now?
- What don't you know now?

4:15 p.m.–4:45 p.m. **Next steps in a policy research agenda**

APPENDIX D
Forum Participants and Their Affiliations

Walter Baer, Senior Policy Analyst, RAND Corporation

John Barclay, Chief Technology Officer, Prometheus Energy Company

Mark Bernstein, Senior Policy Analyst, RAND Corporation

Robert Boehm, Distinguished Professor, Mechanical Engineering, and Director, Center for Energy Research, University of Nevada, Las Vegas

Hazen Burford, Vice President, Operations, Intelligent Energy

Michael Canes, Director, LMI Research Institute

Steve Chalk, Hydrogen Program Manager, U.S. Department of Energy

Andres Cloumann, Marketing Director, Electrolysers, Norsk Hydro Electrolysers AS

Tama Copeman, Director, Future Energy Solutions, Air Products and Chemicals, Inc.

Gary Dixon, Manager, Special Assignments, South Coast Air Quality Management District

Lloyd Dixon, Senior Economist, RAND Corporation

Ronald A. Friesen, Executive Director, Stationary Fuel Cell Collaborative, California Air Resources Board

Devinder Garewal, Director, Strategy Development and External Affairs Office of Chairman, California Air Resources Board

Allan Grant, Manager, Hydrogen and Fuel Cell Program, BC Hydro

Jay Griffin, Doctoral Fellow, Pardee RAND Graduate School

Thomas J. Gross, United States Naval Reserve (RADM, Retired); Associate, IF, LLC

David Haberman, IF, LLC

Nanci Haberman, IF, LLC

Barbara Heydorn, Senior Consultant, SRI Consulting Business Intelligence

Ray Hobbs, Chief Engineer, Future Fuels Program, Arizona Public Service

Karen Kimball, Vice President, Parsons Corporation

Aaron Kofner, Associate Quantitative Analyst, RAND Corporation

Stephen Kukucha, Director, External Affairs, Ballard Power Systems

Robert Lempert, Senior Physical Scientist, RAND Corporation

Sergej Mahnovski, Doctoral Fellow, Pardee RAND Graduate School

Marissa Mittelstadt, Administrative & Logistics Coordinator, Advanced Technologies Group, Toyota

Matt Miyasato, Technology Demonstration Manager, South Coast Air Quality Management District

Charles A. Myers, Director of Sales, Nuvera Fuel Cells

Catherine Padro, Project Leader, Hydrogen Systems, Los Alamos National Laboratory

Geoffrey Partain, Manager, TMS Environmental Vehicles Group, Toyota

D. J. Peterson, Associate Policy Analyst, RAND Corporation

Bill Reinert, National Manager, Advanced Technologies Group, Toyota Motor Corporation

Jim Reinsch, Senior Vice President, Bechtel Power Corp; President, Bechtel Nuclear

Douglas M. Rode, Principal and Managing Director, Hydrogen Safety, LLC

Gerry Runte, Applied Research and Engineering Sciences (formerly Executive Director, Hydrogen Energy Systems Center, Gas Technology Institute)

Maxine Savitz, Honeywell (retired)

Paul Scott, Chief Scientific Officer, ISE Research

Jon Slangerup, CEO, Solar Integrated Technologies (formerly President and CEO, Stuart Energy)

George Sverdrup, Technology Manager, Hydrogen, Fuel Cells, and Infrastructure Technologies, National Renewable Energy Laboratory

Alfred Unione, Santa Fe Operations Manager, Applied Research and Engineering Sciences

Nicholas Vanderborgh, Advisor, South Coast Air Quality Management District

Cynthia Verdugo-Peralta, Board of Governors, South Coast Air Quality Management District

Rick Zalesky, President, Hydrogen Business, Chevron Corp.

APPENDIX E

Matrices Used in the Exercise Described in Chapter Four

Figures E.1 through E.3 illustrate the three matrices used in the exercise described in Chapter Four. Each figure displays the impact of one of three approaches to hydrogen policy—market-only, moderate, and aggressive—and the level of impact for various investment and policy goals given the four future scenarios described in Chapter Four. Forum groups color-coded the matrices to indicate their ideas about likely outcomes. The matrices are reproduced here in grayscale.

Figure E.1
Impact of Market-Only Policy Approach

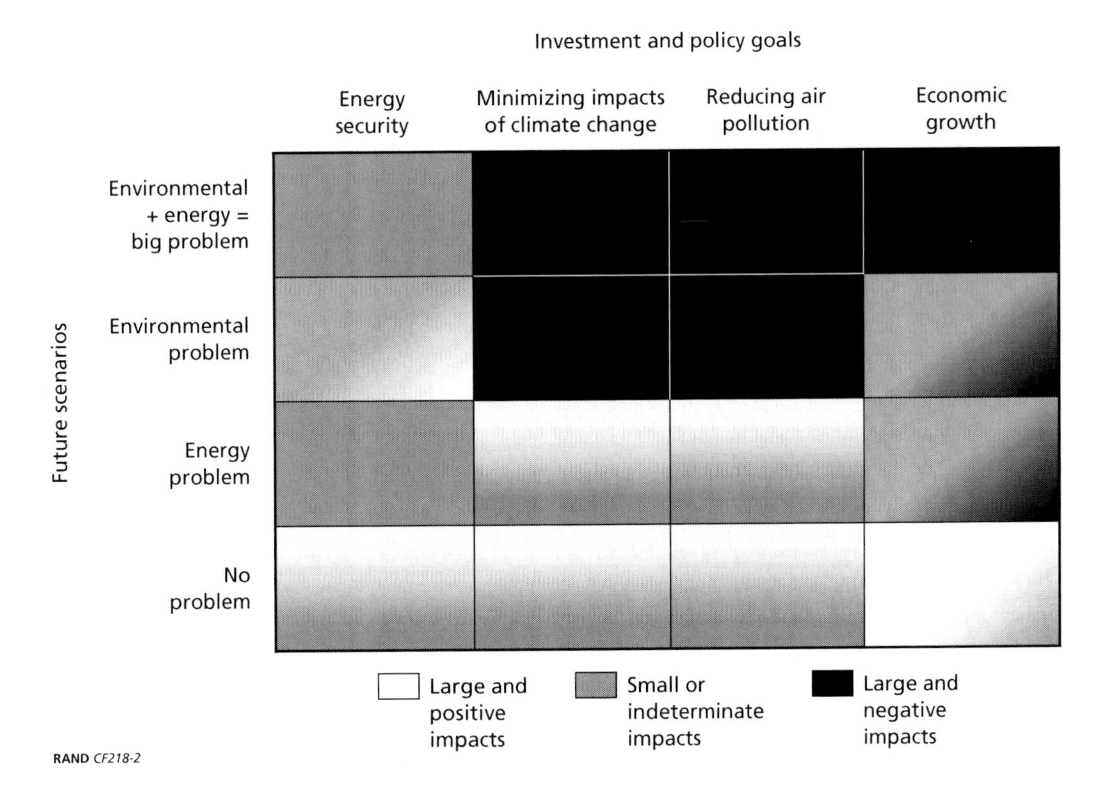

RAND *CF218-2*

Figure E.2
Impact of Moderate Policy Approach

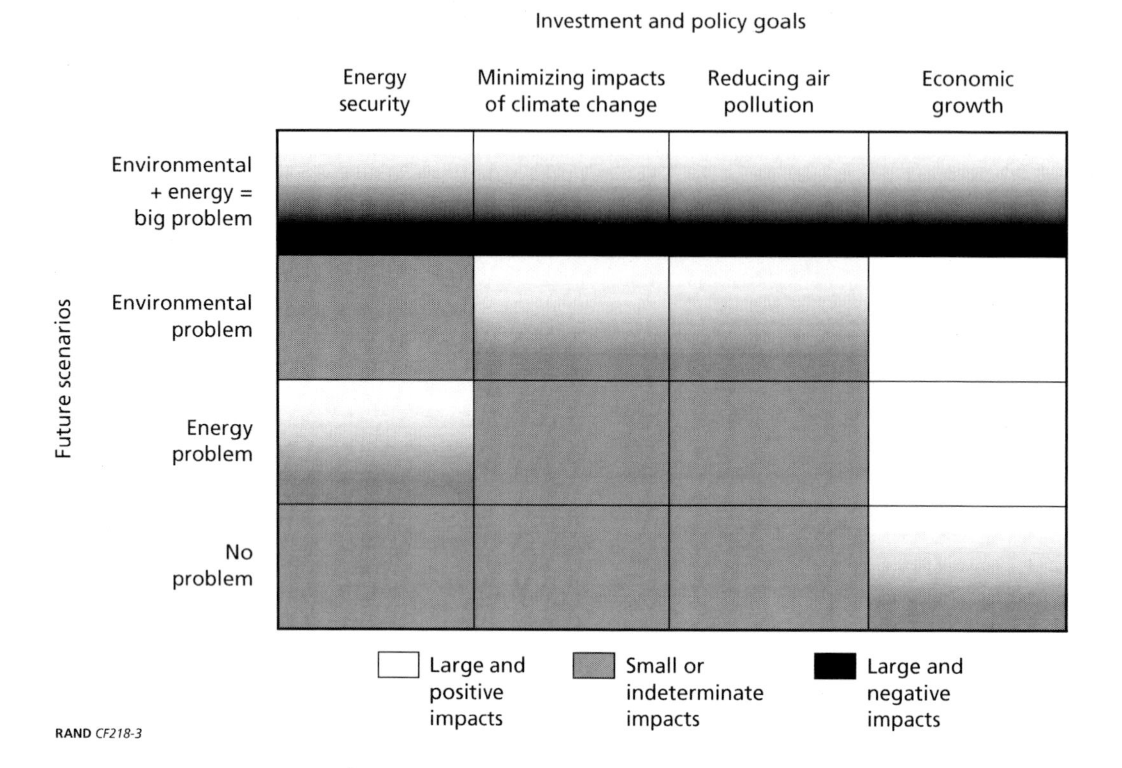

Figure E.3
Impact of Aggressive Policy Approach

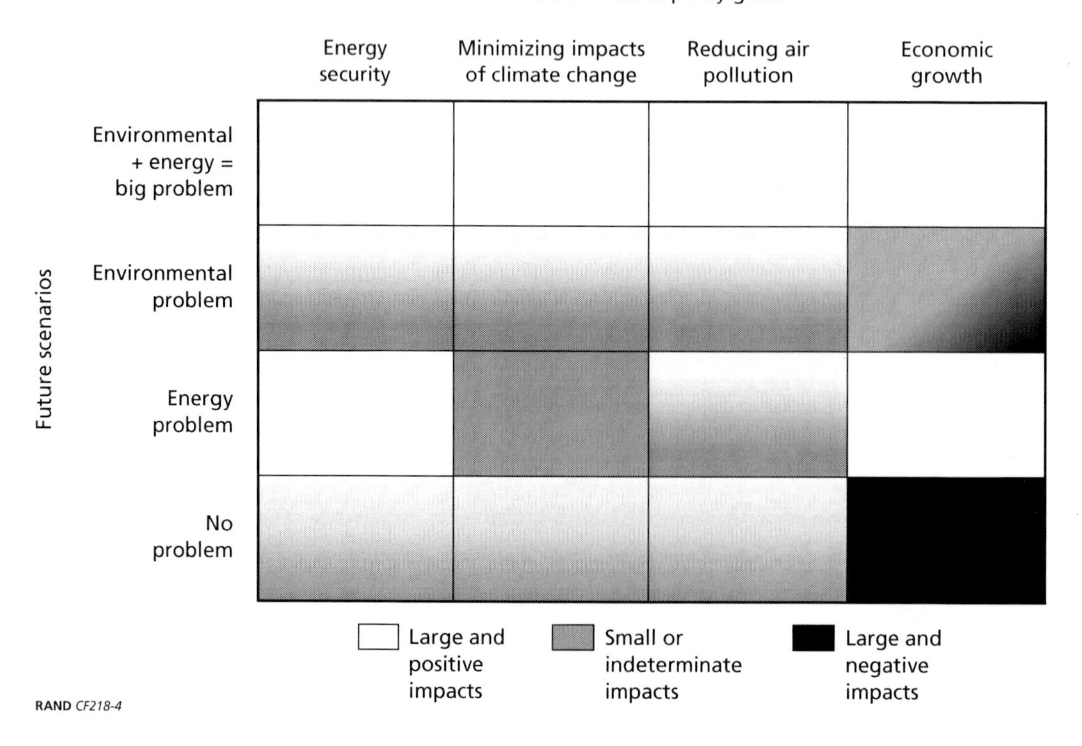